Edward Heawood

Geography of Africa

Edward Heawood

Geography of Africa

ISBN/EAN: 9783337121518

Printed in Europe, USA, Canada, Australia, Japan

Cover: Foto ©berggeist007 / pixelio.de

More available books at **www.hansebooks.com**

OF
AFRICA

BY

EDWARD HEAWOOD, M.A.

FELLOW OF THE ROYAL GEOGRAPHICAL SOCIETY

WITH ILLUSTRATIONS

London
MACMILLAN AND CO., Limited
NEW YORK: THE MACMILLAN COMPANY
1896

PREFACE

In writing this text-book two main principles have, as far as possible, been kept in view throughout. In the first place, the rule laid down by Dr. Mill in the *General Geography* of this series, of proceeding from the general to the particular, has been adhered to; and in the second, a clear understanding of the broad physical features of each region described has been taken as the necessary basis on which to build up the complete picture of such region as the sphere of human activity. Thus the first four chapters are devoted to a consideration of the broad facts which apply to the continent as a whole; and in these, as in each subsequent chapter, the following order has been observed. The physical description starts from the more general characteristics of form, structure, and relief, the hydrographical features to which these give rise being next dealt with, after which a brief description of the chief climatic influences at work prepares the way for the account of the vegetable and animal life. The general nature of the country having been thus brought out, the human inhabitants are considered in their relation to their surroundings, beginning from the native races, or those settled in the country for long ages, proceeding to the more recent relations which have sprung up with the rest of the world, and concluding with the present political status of the various portions of the continent.

As regards the subdivision of Africa into regions suitable for consideration in the separate chapters, unusual difficulties present themselves. In the case of other continents the political units have in course of time acquired so much individuality, and owe their existence so far to causes connected with the configuration of the surface, that they supply the natural, and in fact necessary, basis for a detailed geographical treatment. But Africa is still in a state of transition. Boundaries are undefined, or have been but newly laid down; while the political units are so confusedly scattered over the surface, and in the absence of sharp natural dividing lines derive their *raison d'être* so little from natural causes, that it would be impossible to start from a political basis without losing sight of the broad physical facts, on which, as already stated, it is considered that the chief stress should be laid. It has therefore seemed well to divide the continent into regions each possessed of a certain natural unity, and at the same time homogeneous to some extent as regards its history and political development, although composed at the present day, it may be, of several separate political units. The physical features and native inhabitants of each region can thus be comprehensively dealt with, while the facts which more particularly relate to the progress and capabilities of the separate units are treated of subsequently under the various political headings.

The facts recorded have been derived from a variety of sources, including the accounts of the best-known travellers, and papers published in scientific and other periodicals, both English and foreign. Of works relating to Africa as a whole and general books of reference the following, in particular, have been consulted: Mr. A. H. Keane's *Africa*, in the new issue of Stanford's Compendium; Dr. W. Sievers's *Afrika*, published by the Leipzig Bibliographical Institute; Mr. Keltie's *Partition of Africa*, and *Statesman's Year-book* for 1896; Mr. A. Silva White's *Development of Africa*, with the series of maps by Mr. Ravenstein; Mr. G. G. Chisholm's *Handbook of Commercial Geography*, and *Longman's Gazetteer*; and Mr. Keane's *Ethnology* in the Cambridge Geographical Series; while for the chapter on the Sahara M. Schirmer's work on that region has been of great assistance, and for South Africa, Professor Wallace's *Farming Industries of Cape Colony*. Professor Supan's sketch of "A Century of Exploration," published in *Petermann's Mitteilungen* for 1888, also deserves special mention. Much care has been taken to avoid inaccuracies, and it is hoped that any which may have crept in are not serious. Information regarding such will be welcomed.

In conclusion, it is a pleasant duty to acknowledge the great assistance received during the preparation of this little work from Dr. H. R. Mill, Librarian of the Royal Geographical Society, who has not only given valuable advice at every turn, but has read through the whole of the proofs, and made many suggestions. Acknowledgments are also due to the various gentlemen who have kindly allowed their photographs to be used for the illustrations, to the Council of the Royal Geographical Society for the loan of the same from the Society's collection for the purpose of reproduction, and to Mr. J. Scott Keltic for the use of two illustrations which have appeared in the *Geographical Journal*.

E. H.

1 SAVILE ROW, LONDON, W.,
November, 1896.

CONTENTS

CHAP.		PAGE
1.	GENERAL SKETCH: PHYSICAL FEATURES	1
2.	RACES OF MAN IN AFRICA	19
3.	EXPLORATION	31
4.	POLITICAL RELATIONS	49
5.	NORTH AFRICA	60
6.	THE SAHARA	76
7.	THE SUDAN	86
8.	THE NILE REGION	107
9.	NORTH-EAST AFRICA	129
10.	EAST AFRICA	143
11.	THE ZAMBESI REGION	163
12.	SOUTH AFRICA	180
13.	WEST CENTRAL AFRICA	207
14.	AFRICAN ISLANDS	227
	SUMMARY	237
	INDEX	251

ILLUSTRATIONS

FIG.		PAGE
1.	CRYSTALLINE ROCKS IN THE WESTERN SUDAN	4
2.	LAKE TANGANYIKA	6
3.	FOREST OF WEST EQUATORIAL AFRICA	13
4.	GIRAFFES ON EAST AFRICAN STEPPE	15
5.	ARABS WITH CAMELS	20
6.	NEGROES AT MARKET, WEST AFRICA	22
7.	WALL AND DITCH OF WEST AFRICAN TOWN	26
8.	RUINED MOSQUE, EAST AFRICA	32
9.	TIMBUKTU	42
10.	RABBA, ON THE NIGER	56
11.	GREAT GATE, MEQUINEZ	67
12.	STREET IN CONSTANTINE	71
13.	SAND-DUNES IN THE SAHARA	79
14.	LAGOON, WITH MANGROVES	91
15.	RIVER ANCOBRA, GOLD COAST	92
16.	KANO	96
17.	CAPE COAST CASTLE	101
18.	THE UPPER NILE	109
19.	GENERAL VIEW OF CAIRO	121
20.	THE PYRAMIDS	125
21.	SCRUB VEGETATION IN SOMALILAND	135
22.	ELEPHANTS IN THE "NYIKA" STEPPE	145
23.	PARK-LIKE SCENERY IN EAST AFRICA	151
24.	BAMBOO FOREST, MOUNT ELGON	152
25.	MOMBASA	159
26.	SCENERY IN EASTERN MASHONALAND	165

FIG.		PAGE
27.	THE VICTORIA FALLS .	167
28.	POST OFFICE AT THE MLANJI STATION	176
29.	MOUNTAIN SCENERY ON THE SOUTH-EAST COAST	183
30.	CAPE TOWN AND TABLE MOUNTAIN .	197
31.	BOTANICAL GARDENS, CAPE TOWN	198
32.	RAPID ON THE ARUWIMI .	212
33.	STREAM IN THE EQUATORIAL FOREST	215
34.	SCENERY NEAR THE LOWER CONGO	223

MAP OF AFRICA .   *To face Preface*

NOTE

In the spelling of native names of places the rules laid down by the Royal Geographical Society have been followed in the main, the chief features being the pronunciation of the vowels as in Italian, and of the consonants as in English. Thus—

    a is pronounced as in *father, amount*
    e ,, ,, *fête, wet*
    i ,, ,, *ravine, hit*
    o ,, ,, English generally
    u ,, ,, *lute, pull*
    ai is pronounced as *i* in *ice*
    au ,, ,, *ou* in *loud*.

With regard to the consonants the chief points to be remembered are that *ch* is always pronounced as in *church*; *g* is always hard; *c* (hard) is replaced by *k*, *qu* by *kw*; *r* is always sounded, and is not used as in English to lengthen a vowel; *s* has always a sharp hissing sound as in *house*, and is never soft like *z*, as in *hose*.

CHAPTER I

GENERAL SKETCH—PHYSICAL FEATURES

IN this chapter we will examine briefly the broad physical features of the continent of Africa, and we shall see later the effect which these have had on the life of man upon it.

1. *Form and Surface*

Position.—Although Africa has been actually an island since the cutting of the Suez Canal through the isthmus which formerly united it with Asia, we must not lay too much stress on this fact; for, on looking at a map, we see that while to the west, south, and east the land-mass is bounded by broad and deep oceans, beneath whose waters it sinks with surprising steepness, on the north and north-west it is separated from Europe and Asia only by comparatively narrow seas, which at certain points are still further narrowed to form the Straits of Gibraltar and Bab-el-Mandeb. On these sides, therefore, it is not so completely divided off as on the others from other land areas, and it may be considered as really part of the great land-mass which is known as the Old World. Again, the Red Sea, which divides Africa from Asia on the north-east, is both narrower and shallower than the Mediterranean, by which it is separated from Europe on the north, so that physically Africa stands in a closer relation to Asia than to Europe. This is also shown by the fact that as regards climate and other characters resulting therefrom, Africa has much more in common with

Asia than with Europe, owing to the similar position of Northern Africa and Southern Asia with regard to the equator. Since the equator cuts Africa at an almost equal distance from its northern and southern extremities, a larger proportion of that continent lies within the tropics than is the case with any other, but owing to the narrowing of the land-mass southwards, a much larger area lies to the north than to the south of the equator.

Outline.—Africa is distinguished among the continents by its remarkable simplicity of outline. Instead of being cut up by arms of the sea extending far into the interior, it may be said to consist of two regularly shaped segments joined at right angles, a northern stretching from east to west and a southern stretching from north to south. The Gulf of Guinea on the west coast merely occupies the angle between these two segments, while the Red Sea on the north-east, which penetrates far into the combined mass of Asia and Africa, does not break the regularity of the outline of either. In the south-east the Mozambique Channel, which now makes an island of Madagascar, may be held to represent a pair of gulfs united by the final separation of the island from the main mass. In the north the gulfs of Sidra and Gabes (or the "Syrtes") form in combination but a single shallow incurve, nearly three times as wide as it is deep, and elsewhere the only irregularities on a large scale consist of the alternation of gentle outcurves and incurves of the coast, with an occasional sharper inlet on a small scale, such as False Bay, near the Cape of Good Hope, and Delagoa Bay in the south-east. As there are no deep bays, there are, of course, no large peninsulas; the nearest approach to one is the portion of the southern limb which runs eastwards to the south of the Red Sea, and has been termed the eastern "horn" of Africa. Bold capes and promontories are not, however, wanting, a large portion of the coast being high and rocky. Such are the Cape of Good Hope; Cape Agulhas, the southern point of the continent; Cape Guardafui, at the extremity of the Eastern Horn, and many points on the north and north-west coasts, as well as on those of Guinea.

The great regularity of outline of the continent is further shown in the absence of large islands lying near the coasts, such as occur so frequently around Europe and Asia. We find instead merely a few groups or single small islands, mostly of volcanic origin, such as the Canaries and Cape Verdes off the north-west coast, Fernando Po and three smaller islands placed in a line at the head of the Gulf of Guinea, and Zanzibar, Pemba, and Sokotra off the east coast. Madagascar is, indeed, a large island, but it is separated by a wide and very deep channel. From the neighbourhood of the equator on the west coast to the Mozambique Channel on the east there is not a single island of any importance in a distance of 4000 miles. Africa thus forms a very compact mass. In area it is only exceeded among the continents by Asia, which contains (with its islands) over 17,000,000 square miles as compared with the 11,500,000 of Africa.

The **structure** of the continent shows to some extent a similar uniformity. Although it is still uncertain of what rocks a large proportion of the surface is composed, we know, at least, that in various places very ancient crystalline rocks are found; in parts these are spread over large areas, but in many cases they form mountain ranges, and may be likened to the framework or skeleton of the whole continent. On these rest a variety of rocks of later date, but, on the whole, there is a striking absence, except quite in the north, of the more recent deposits formed on the floors of ancient seas. Yet large parts of the surface are covered with sands or alluvial soil, deposited either by the winds or by rivers, or on the floors of ancient inland lakes. Another recent formation which covers large areas in West Africa is a yellowish or reddish clay, known as laterite, formed from the decomposition of the rocks by the action of the air. It becomes baked by the heat of the sun into very hard masses. In many places, but principally within a tract of country stretching from north to south in the eastern part of the continent, we find vast quantities of rocks, which have, at a comparatively recent date, been poured forth from the interior of the earth by volcanic agencies. From the absence of the newer rocks formed beneath the sea, it is fair to conclude that Africa as a whole is a very ancient continent, and that the greater part has remained since very remote times above the surface of the water.

Relief.—Another marked characteristic of Africa is its uniformly high level. Although the highest elevations are by no means very great as compared with those of other continents, it is second only to Asia as regards its average height, which has been calculated to be about

FIG. 1.—CRYSTALLINE ROCKS IN THE WESTERN SUDAN.
(Photograph by Rev. J. T. F. Halligey.)

2000 feet. We nowhere see extensive plains slightly raised above the level of the sea, such as the vast low plains of Northern Asia or of the Amazon valley in South America. Almost everywhere there is a sudden rise at no great distance from the coasts, and it is only where rivers have worn down for themselves channels through the higher ground that the low land stretches any distance into the interior. As a rule, on penetrating inland from the coast one or more steep ascents are soon reached, leading either over mountain ranges parallel to the coast, or up a series of terraced escarpments to a high plateau which fills up the greater part of the interior. On this account Africa has been likened to an inverted saucer, but this comparison rather holds good for certain sections of the continent, especially in the southern half, than for the whole taken together. In the first place, the whole of the southern (or north to south) limb has a much higher average elevation than the northern (or east to west) limb, which supplies a second point of contrast between the two. Besides this, the elevated rim is not nearly so continuous in the northern as in the southern section, being almost wanting to the west.

In the extreme north-west the isolated mountain range of the Atlas, with peaks 13,000 to 14,000 feet above the sea, runs mainly east and west parallel to the north coast. Apart from this, the principal lines of elevation of the northern plateau consist of long arms of high ground, which branch off from the higher southern plateau in various directions. One of these runs north and south, close to the east coast, a second stretches right across the centre of the plateau from south-east to north-west, and a third (separated, however, from the southern plateau by the lower course of the Niger) runs east and west parallel to the shore of the Gulf of Guinea. Between these lie stretches of lower ground, the most central forming an altogether closed basin, bounded on all sides by higher land, and dipping to Lake Chad.

The higher southern plateau may be broadly subdivided into a higher eastern and a lower western division; the former contains almost all the highest mountains of the continent, except the Atlas range and the isolated peak of Cameroons at the angle between the northern and southern limbs. It is marked by abundant evidences of volcanic activity, consisting on the one hand of a number of volcanoes (for the most part extinct), and on the other of large areas covered with old lava-flows. Unlike most volcanoes, those of Africa are not as a rule found in the vicinity of the sea, some of them being situated at a distance of at least 700 miles from the nearest point on the coast. These old volcanoes are the highest mountains in Africa. The most important (which rise above the limit of perpetual snow) are Kilima Njaro and Kenya, near the eastern edge of the plateau, and Ruwenzori, in the very centre of the continent, near the Nile sources. Each is about 18,000 to 19,000 feet high. Great earth-movements of a different kind are also indicated by the presence of a system of longitudinal valleys with steep walls, running nearly due north and south in an almost unbroken line

FIG. 2.—LAKE TANGANYIKA.
(After Baumann.)

from the southern parts of the continent up to the Red Sea, and continued farther north in the Dead Sea and Jordan valley. It has been suggested that this great line of depression has been formed by a vast crack or sinking of a portion of the earth's crust. An important characteristic of this rift in the elevated eastern plateau, and of the plateau itself, is the presence of a number of large lakes, in such numbers and of such a size as occur nowhere else on the surface of the globe except in North America. These lakes vary greatly in size and shape, but may be divided into the broad, and the long narrow types. The narrow lakes occupy portions of the great valley system just referred to, and are, as a rule, exceedingly deep, and their shores are higher and more precipitous than those of the other type.

The western part of the southern limb of Africa, though retaining on the whole the form of a plateau, contains an enormous circular basin of very gentle slope, and almost enclosed by higher ground, which must have been formerly occupied by a vast inland sea, and is now drained by the river Congo and its tributaries. This depression is almost bisected by the equator, and occurs just opposite that part of the eastern section which contains the most elevated ground of all, so that it is here that the contrast between east and west is sharpest. Farther south there is much greater uniformity of level, but there too the Drakensberg, the principal range of South Africa, keeps close to the south-east coast.

River Systems.—Although the whole of the interior of the continent has been spoken of as a plateau or table-land, this must be understood only in a very general sense, for when we examine the river systems, which supply the best indication in detail of the slope of the land, we find that there is a good deal of variation in different parts. The rivers of Africa may be broadly divided into two classes, on the one hand those which rise on the outer edge of the main mass of high ground, and run directly down to the sea, and on the other those which rise on the inner side of the ranges by which the plateau is bounded, and either find their way by a long and devious course to the sea, which they often reach on the opposite side of the continent, or empty themselves into inland lakes. In order to reach the sea it has been necessary for them, in the course of long ages, to cut a more or less narrow passage through the bounding ranges on the side remote from their source, and these passages are, as a rule, obstructed by rapids or cataracts. The rivers of the former class have comparatively short courses, and, except when in flood, do not carry down a large volume of water. Those of the second class are naturally much more important, and their combined basins or drainage areas cover the greater part of the continent. The four principal rivers of the continent collect most of the rainfall which reaches the sea, while the total number of rivers in this class deserving mention does not exceed seven.

The irregularity of the slope of the continent may be seen from the varied directions taken by these rivers. Beginning with the **Nile** (the longest river of the continent, and traversing a greater extent of country from source to mouth than any other river in the world), we see that the main slope of its basin is due north. Its sources lie on the higher southern plateau, but it soon descends with an abrupt fall to the lower level in the north, continuing to receive supplies of water from the high lands to the east, which separate the basin of the river from the Red Sea throughout their whole length. To the west the Nile basin is bounded by few well-defined ranges of mountains, and during the lower half of its course the country in this direction is quite low. As there are no marked ranges of hills, and we are still very ignorant of the slope of the country, the precise limits of the basin cannot be drawn on the map, but as the whole country is devoid of running streams, it is limited, as far as water supply goes, to the actual valley of the stream; the broadest part of the basin thus occurs in the upper half, its width at about 10° north of the equator being over 1100 miles.

West of the Nile basin occurs that of **Lake Chad**, enclosed, so far as we know, on all sides by higher ground, so that none of its waters ever reach the sea. Still farther west, we come to the basin of the **Niger**, with a main slope from north to south, or just the reverse of that of the Nile, although the principal stream starts in a direction almost opposite, and describes an enormous curve, nearly a semicircle, before reaching the sea. Like the western limit of the lower Nile basin, the northern limit of the Niger basin is undefined, and for the same reason, but it appears that a large arid area to the north, from which no water now comes, properly belongs to the basin of the river. The southern limit is formed to the west by a narrow strip of plateau parallel to the coast of the Gulf of Guinea, rising to a considerable height only at the western extremity. From the east comes the largest tributary, the Benue, which has its source on the high table-land of the southern section of the continent.

The first river-basin of this high southern table-land which we come to from the north is that of the **Congo**, the most important of all in every respect except the mere length of the main stream; it occupies a vast, nearly circular area in West Africa, with a general slope from east to west, touching on the north each of the three basins already mentioned. The outer parts on all sides consist of high plateau-lands, a large portion to the east belonging to the high eastern table-lands, and reaching to within a comparatively short distance of the east coast. From all sides, however, the ground slopes downwards to the less elevated central depression, which appears to be the dried-up bed of an ancient inland sea. The combined waters of this vast area finally break through the western rim of the continent by a narrow passage, in which the stream is constantly broken by rapids and cataracts, and becomes placid once more only when within a short distance of the sea.

South of the Congo basin we come to that of the **Zambesi**, which also stretches across almost the whole width of the continent, but in its general slope is exactly the reverse of the Congo, flowing as it does from west to east. The elevation of the basin of the Zambesi is both great in itself and remarkably uniform, and only along the lower course of the main stream and its chief tributaries is there a decided break in the general altitude of the country. On the whole, the limits are not marked by important ranges; in fact, there are parts of the water-parting on the side of the Congo where the line of partition is hardly distinguishable, so level is the country, while on the south there is a considerable interlacing of its streams with those connected with the small Lake Ngami, the direction of some of them even varying with the time of year. Towards the east the boundaries are better marked, and the stream has, like the Congo, to break through the rim of the continent before reaching the sea.

The two remaining river-basins which occupy South Africa are of much smaller size. That of the **Limpopo** touches the Zambesi basin on the south-east, and it has a larger proportion of its area on the seaward face of the plateau rim. In the case of the **Orange** the direction of slope is again reversed, the head-waters being divided from the east coast only by the elevated Drakensberg Mountains, while its lower course breaks through to the sea on the west. The basin extends much farther to the north than to the south of the main stream, but owing to the dryness of the country the river beds in this direction are empty during the greater part of the year.

These main river-basins, with the coast streams near them, form in some ways the most natural subdivisions of the continent. But it is not possible to divide up the land into distinct divisions with reference to the river systems alone. The high lands which may form part of several river-basins are also natural regions of great importance; a notable example of this is the high lake-plateau of East Africa.

2. *Climate, Flora, and Fauna*

Climatic Contrast.—No one factor perhaps has such a potent effect on the general condition of a country and on the life of its inhabitants as climate. In this respect also we find a contrast between the northern and southern divisions of Africa, although the boundary line does not occur in exactly the same position. The most important point of contrast consists in the fact that the greater part of North Africa has a very scanty rainfall, while, on the whole, the southern section has an abundant supply. It is true that in the south a similar region of scanty rainfall occurs at about the same distance from the equator as in North Africa, but, owing to the shape of the continent, which becomes narrower the farther we proceed in this direction, this region is limited in extent, and exercises but a slight influence on the southern half of the continent.

Causes of the Desert.—This climatic contrast has

such an important bearing on the whole character and history of the regions in question, that it will be well briefly to glance at some of the causes which bring it about. The distribution of rain is due principally to the winds which prevail, and these again are very largely influenced by the distribution of land and water and the general configuration of the ground. It is easy to see that, owing to the shape of the continent and the east to west direction of the northern limb, a much larger proportion of land falls within any given latitudes to the north than to the south of the equator, and the effect of this is heightened by the proximity of the land-mass of Asia, separated only by a comparatively narrow sea. Northern Africa and South-West Asia, in fact, form part of one climatic region, a study of meteorological maps showing us that all the most important climatic conditions which prevail on one side of the Red Sea are exactly reproduced on the other. The effect which this great mass of land in North Africa, withdrawn in large measure from the influence of the sea breezes, has in making the climate different from that of Central and Southern Africa may be seen in various ways. Partly on this account, and partly because of the lower elevation, a great part of Northern Africa is subject to an intense heat, which is unknown farther south, and generally the climate is what is known as an extreme continental one, the difference in temperature between the hottest and coldest months and between day and night being much greater than in the other parts of Africa. There is also a greater variation in the barometric pressure at different seasons, and the mean for the year is much lower. The precise manner in which these conditions act on the winds, and through them on the rainfall, cannot be explained in few words, but the general result is that while for a portion of the year Northern Africa forms a centre whence currents of air diverge in all directions, during the remainder, when winds blow in from the surrounding regions, they are either scantily charged with moisture at the outset, or lose the greater part of it in passing over mountain ranges like the Atlas at the outer edge of the region. The air also is made hotter, and so tends to evaporate more water, rather than to deposit any as rain.

Minor Climatic Areas.—The contrast here drawn, of course, only holds good as a broad generalisation, for in the extreme north, as in the extreme south of the continent, there is a region possessed of a comparatively temperate climate. Along the coast of the Mediterranean, and especially north of the line formed by the Atlas range, the climate generally resembles that of the opposite or European shore of the Mediterranean, both as regards the mean temperature of the year and the annual amount of rainfall. Again, in the southern half of the continent the low-lying coast-lands on the east have a high temperature.

Tropical Africa.—On the whole, it may be said that the temperature of Africa south of the northern deserts is low considering the latitude, this being due principally to the great average elevation. On the higher plateau, in fact, the climate is often really cold. There is also much less difference between the hottest and coldest months there than in North Africa. In the southern half of Africa south-west winds are the most prevalent on the west coast, owing to the generally high pressure of the air over the Atlantic. On the east coast there is an alternation between the north-east and south-west monsoon. The north-east monsoon blows during the summer of the southern hemisphere, when the pressure is low over Central and South Africa, but high over South Asia and the Indian Ocean. This lasts from December to March. From April to October the south-west monsoon blows. In the Red Sea the wind is deflected, and the monsoons become south-east and north-west respectively. On the south-east coast of Africa the south-east Trade is the most usual wind. The rainfall, though fairly abundant everywhere in the southern half of Africa, never reaches the excessive proportions which it does in some other parts of the world, as *e.g.* in the districts at the head of the Bay of Bengal. There is considerable variation in the distribution of the rainfall through the year. On and near the equator there are two rainy seasons separated by a short dry one, but as the distance from the equator increases only one rainy season is found, the remainder of the year being continuously dry. In the extreme south the condi-

Fig. 3.—Forest of West Equatorial Africa. Banana in Foreground. (After Stuhlmann.)

tions are different, being those belonging to the temperate zone.

In spite of the absence of excessive heat in Equatorial Africa, the relatively large amount of moisture in the air makes it very trying to Europeans. The damper regions, especially the coast-lands, are particularly subject to malarial fevers, which prove very fatal to the inhabitants of temperate climates, and make it doubtful whether Europeans can ever settle permanently in the country. The more elevated central plateaux are, however, more healthy, while the desert regions, in spite of their excessive heat, are almost as healthy as the northern and southern temperate districts.

Vegetation.—The vegetation of a region, which has such an important influence both on its external aspect and its capabilities is, in its turn, mainly dependent on climate. We may therefore expect to find great contrasts in Africa in this respect also. Above all, an abundant rainfall helps the growth of plants, while differences in the flora of different regions result from variations of temperature, and of its distribution through the year. Thus the temperate regions of the extreme north and south possess special floras quite different from that of the tropical portion of the continent. That of the south consists largely of shrubs and bushes, the most characteristic order of plants being the heaths (*Ericaceae*), which occur in surprising variety. In the desert regions, especially in the north, the growth of plants is limited, and in the most arid parts the flora consists of a few species only, which in course of time have become specially adapted to an excessively dry climate. In all the dry parts of the continent species of mimosa and acacias are particularly abundant.

Tropical Types.—Even in the regions which are favoured with a fairly plentiful rainfall great differences are to be observed. The growth of true tropical forest, matted together with creepers and choked with dense undergrowth, requires the combination of heat and rainfall distributed pretty evenly through the year. Such conditions are found principally on or near the equator, where there is no lengthened dry season, and, within this zone, mostly in the western half of the continent, where the temperature is higher on the whole and more equable than on the high uplands of the eastern half. It is, therefore, in West Central Africa, in the Congo basin and on the coast-lands of the Gulf of Guinea, that we find the most extensive and unbroken forests, which, though hardly anywhere attaining the rich luxuriance of those of South America, still cover large expanses, especially in the inner parts of the Congo basin. This western region is also the home of two species of palms, which are mainly confined to it, viz. the oil and wine palms (*Elaeis guiniensis* and *Raphia vinifera*).

By far the larger part of the surface of Tropical Africa is, however, covered with wide grassy expanses, either

Fig. 4.—Giraffes on East African Steppe.
(After a drawing by W. Kuhnert.)

with or without trees; this grass often attains a vast height unknown in temperate climates. Frequently the courses of streams are fringed continuously with narrow belts of forest, to which the name of gallery forests has been applied, while away from the moisture of the stream the country is almost bare of trees. This fact has often caused the extent of forest to be over-estimated, for, as explorers have very frequently followed the courses of streams for long distances without diverging on either side, they have gained the impression that the whole country they have passed through has been forest-clad. A still more characteristic form of vegetation (perhaps most of all characteristic of Africa as a whole) is that of the tree savannahs, in which the grassy expanses are dotted over with trees either singly or in small patches, giving the country a park-like appearance. Very characteristic of these savannahs are the curious tree-euphorbias, which, from their shape, have been likened to candelabra. The grass is, as a rule, burnt down annually by the natives, in order that a fresh and tender crop may spring up in its place; and these annual grass-burnings, whose smoke fills the whole air with haze for weeks together, have doubtless done much in the past to limit the extent of forest by thinning the growth of trees at the edges. Where forest is found in the eastern half of the continent it presents, as a rule, a very different appearance to the moist forests of West Africa, the trees being stunted and thin, and creepers and undergrowth almost absent, while in the dry season the whole is so scorched up as to present a mournful and almost lifeless appearance. In some of the drier parts of East Africa trees are almost entirely wanting, but grassy steppes extend for miles, forming grazing-ground for cattle or for vast herds of wild animals. (See Fig. 4.)

In some of the mountainous regions near the coast, however, luxuriant forests cover the upper regions, which condense the moisture from the sea breezes, and from affinities which these possess with West African forests, it has been thought that they once extended much farther east, when the climatic conditions were more favourable to their growth.

Other types of vegetation are : (1) that which fringes the margins of swampy lakes or streams with slight current, especially on the Upper Nile. It is characterised, above all, by the presence of papyrus growing at the edge of the water; (2) the mountain vegetation of the higher East African peaks and the table-land of Abyssinia, which recent researches have shown to possess a striking unity, though at present occurring in isolated spots far removed from each other.

General Characteristics.—As compared with other tropical countries, Africa has not a very varied flora, and in particular has but few species of palms. One of the most peculiar and widely distributed trees is the baobab, marked by its gigantic trunk and branches, but scanty foliage. It is found everywhere in the central zone. The coasts, where low and swampy, are everywhere fringed with mangroves, forming impenetrable thickets, and in slightly drier spots the *Pandanus*, with its whorls of spiny leaves, is generally to be seen. Some of the most useful plants found in Africa in a wild or semi-wild state may be here enumerated. The chief are the date palm, esparto grass or alfa (used for making paper), and various gum-acacias and aromatic plants in the dry regions of the north; the banana in all the moister and warmer parts; the oil palm, orchilla weed (used as a dye), and numerous creepers yielding indiarubber, in the moist forest regions; and the cotton plant and coffee shrub, the kola nut and the gum-copal tree in other parts of Tropical Africa. The plants regularly cultivated will be noticed in speaking of the occupations of the people.

Faunal Regions.—Africa is the home of an unusual number of large wild animals, for which the extensive grassy or arid steppes of East and South Africa are particularly favourable. The characteristic fauna of Africa is found almost entirely to the south of the Tropic of Cancer, the northern strip of the continent belonging in this respect to the same region as Southern Europe. A marked contrast is to be noticed between the forest-clad and more open regions of Africa, the former being as poor in animal life as the other parts are rich in it. Travellers state that they have journeyed for days through the forest region of West Africa without seeing a trace of the larger animals. This region (including the greater part of the Congo basin and the shores of the Gulf of Guinea) is, however, the particular home of the anthropoid apes, the fierce gorilla and the chimpanzee, and other special forms of animal life, so that in this respect, as in other ways, it forms a natural subdivision of the continent. In all the forest area cattle-rearing is little practised.

Characteristic Fauna.—Apart from these local differences the fauna of the continent, as a whole, is strikingly uniform, most of the large animals being found over the greater part of its extent, though of late years they have in many parts been driven back by man into the more remote districts, so that their range is yearly becoming more limited.

An exception to this uniform distribution is formed by that of the elephant and camel, the former not advancing into the desert regions of the north, while the latter hardly penetrates southward beyond them. The African elephant is of a distinct species from that found in Asia, a principal difference being in the form of the ears (which are much larger in the African) and in the outline of the skull. The other characteristic animals are : of carnivores, the lion and leopard (though not the tiger), hyaena, jackal, etc. ; many varieties of antelopes, which are often seen together with the giraffe, the zebra, and the buffalo in immense herds on the open steppes, especially in the neighbourhood of water ; the hippopotamus and four kinds of rhinoceros ; of reptiles, the crocodile ; while the ostrich, the largest of existing birds, roams over the more arid regions.

The most striking peculiarities of the African fauna as compared with that of Asia consist in the complete absence of the tiger ; in the almost entire absence of deer, in spite of the many species of antelopes present ; and in the presence of the giraffe and hippopotamus, which do not extend to Asia. The West African region shows remarkable resemblances with the east of Asia, although the farthest removed from it.

CHAPTER II

THE RACES OF MAN IN AFRICA

1. *Physical Characters*

Two Main Types.—The contrast which we have already noticed between North and South Africa, looked at in their broad outlines, may be observed also when we come to study the various races which people the continent. Although all the inhabitants are more or less dark of skin, the resemblance between the dwellers in the northern and southern divisions of the continent ends here, and in all other respects, such as the shape of the skull, the hair, the character of the features, and also in the temperament and mental qualities no less than in the language spoken, marked differences can be seen. The Northern and the Southern Africans belong, in fact, respectively to two out of the three main groups into which the human race has been divided, the North Africans, like almost all the inhabitants of Europe and South-Western Asia, forming part of the so-called White, Caucasian, or Mediterranean type, while the latter belong to what has been styled the Ethiopic or black type. The rough boundary line between the two corresponds, on the whole, remarkably well with the principal climatic line of division of the continent, following in the main the southern edge of the desert regions of North Africa, though we shall see that the northern races have in many cases overstepped this line and mingled with the black race to the south. Even in regard to the colour of the skin the two groups show great differences, for while that of the Southern is black in varying degrees of intensity, the Northern is at most of a brown or dusky hue, and in many instances approaches the paleness of Europeans. In other respects the main characteristics of the Northern and Southern races may be briefly described as follows: the Northern have broad skulls, straight noses, with a high bridge, thinnish lips, and somewhat curly hair; while the Southern have narrow skulls (long in

Fig. 5.—Arabs with Camels.

comparison from front to back), broad, flat noses, thick lips, woolly hair, high cheek bones, and in particular very projecting jaws. These characteristics are to be seen where the races have best maintained their individuality, but it is not uncommon to find among a portion of the black peoples individuals with finely formed features, due, it may be supposed, to an admixture which has taken place in the past with the finer Northern races.

The Northern Races.—These, in their turn, fall under two main divisions of the Mediterranean race, which have been named Semitic and Hamitic. Although the general resemblance between the two is sufficient to make it

evident that in remote ages there must have been a close
connection between them, certain broad differences of
feature can be distinguished. The Hamites have not,
as a rule, quite such high aquiline noses or thin lips or
so much beard as the Semites. Both light and dark shades
of complexion are seen among both races. In Africa the
two races are much mixed up, and their representatives do
not occupy well-defined areas. A great mixture of blood
between them seems to have occurred. The Arab tribes
who, with the Abyssinians, represent the Semites, are found
all along the northern sea-board, but with a large admixture
of Hamites in the north-west. They also occupy the extreme west of the Sahara, and spread southwards in the
Nile valley. In no way is the close relation which exists
between Northern Africa and South-Western Asia shown
more clearly than by the distribution of the inhabitants,
for Arab tribes occupy both shores of the Red Sea, and
even in historic times migrations are known to have taken
place between the two continents. The principal groups of
the Hamites are the Fellahin or present country population of Egypt; the Berbers of the north-west; the Tuareg,
who occupy the west centre of North Africa; the Bisharin
or Beja, between the Nile and the Red Sea; and the
Gallas and Somalis of the eastern horn of the continent.
Probably another branch of the Hamites is formed by the
Fulbe, who, towards the west, have mingled with and
lorded it over the black race to the south of the desert,
but whose kinship has not yet been satisfactorily determined.

The Southern Races.—Almost the whole of the rest
of Africa is occupied by the two main branches of the
black or Negro race. According to the names commonly
used, the first division of this race is described by the
term "Sudan Negroes," from the region which they inhabit,
and which extends between the desert and the Gulf of
Guinea, with a corresponding zone eastwards as far as the
region of the Upper Nile. These tribes, though speaking
a great variety of languages, are all remarkably alike
physically, and show in the highest degree the characteristics already mentioned as belonging to the Negro race.

The tribes of the second branch, known as the Bantu Negroes, live to the south of the Sudan, and occupy almost the whole of the southern limb of the continent. Among them much greater deviations from the true Negro type are found, and this is supposed to be due to the fact that they have in course of time mingled to some extent with the Northern races and acquired some of their characters. All the Bantu tribes speak very nearly allied languages,

Fig. 6.—Negroes at Market, West Africa.
(Photograph by Rev. J. T. F. Halligey.)

different from those of the Sudan Negroes, and this fact forms the main reason for grouping them as a separate branch of the Negro race. It has been suggested that these two divisions correspond to two different migrations in prehistoric times, but this is a mere conjecture, as the origin and early migrations of the Negroes are lost in uncertainty. The fact that similar black races are found in South Asia and Australia lends some support to the idea that such races once occupied a large part of the tropical regions of the Old World, but were gradually pushed southwards by other races. At any rate, if the

Negroes came originally from Asia, their connection with that continent must have been very much more ancient than that of the Northern races, who appear to have crossed over later, and to have encroached on the Negroes from the north. In fact, this process has gone on to some extent down to recent times, especially in East Africa, where tribes of supposed Hamitic origin have been for some time pressing on the Negroes from the north, and where even some of the Negro countries are ruled over by a superior race, said to have come originally from that direction.

In various parts of the Bantu domain scattered tribes of small stature (called dwarfs or pygmies) have been met with in recent years, and are supposed to represent similar tribes spoken of by the ancient classical authors. They differ in many ways from their Bantu neighbours, and as they occur chiefly in the forest regions of the west, it has been conjectured that they form the last remnants of an ancient aboriginal population of Africa. If this were the case, the forest region would be naturally the one in which they would linger longest, screened from the attacks of their more powerful neighbours. It may be, however, that they represent merely Negro tribes who have degenerated in course of time, owing to the unfavourable surroundings in which they are placed.

In the south-west of the continent we find other tribes, totally distinct from the Bantu Negroes, which are classed under the general designations Hottentots and Bushmen. The former name was given to the natives of the Cape of Good Hope by the first Dutch settlers, in ridicule of the peculiar clicking sound of their language. The name they apply to themselves is that of Khoi-Khoin ("the men"). The kinship of these tribes has not yet been satisfactorily determined. The language of the Hottentots possesses some characteristics which indicate a high development. That of the Bushmen, however, though possessing the same clicking sounds, is said to have no real connection with the Hottentot language as regards structure. Both races are of a lighter colour than the Negroes, being of a yellowish brown instead of black, but the Hottentots are decidedly taller than the Bushmen, who rank among the smallest races of the continent. In fact, it has been supposed by some that they are allied to the races of small stature scattered over West Central Africa, and with them form the remnants of an aboriginal population spread over the whole of Southern Africa before the arrival of the Bantus.

2. *Manners and Customs*

Modes of Life.—The mode of life of the various African races differs greatly in accordance with the differences in the nature of the country in which they dwell. There is a broad distinction between the Northern and Southern races in this respect also, owing, doubtless, to the fact that fixed habits have been acquired in course of time during residence in regions particularly adapted to supply the means of livelihood in one manner or another. We have seen that the Semitic and Hamitic races dwell principally in the arid regions of the north, which, though possessing little land suitable for cultivation, afford scanty pasturage for flocks. Like their kindred, therefore, in South-West Asia, who inhabit similar arid regions, these races are in the main pastoral, possessing also the nomadic habits entailed by wandering in search of pasturage. The steppes of East Africa are equally suitable for cattle-rearing, and this helps to explain the great southward extension of the Hamitic race in that region. The composition of the flocks varies in different regions according to the amount of dryness, only the camel being able to subsist in the most arid zone of all. The owners of camels find a special occupation in the provision of transport for the caravans which cross the desert from north to south, while in the oases a more stationary population occupied with cultivation is found.

The Negro race generally practises agriculture, often to the entire exclusion of cattle-rearing, especially in the moist and warm regions of West Africa; the Negroes therefore, as a rule, are more stationary in their habits.

In the south of their domain, however, the nature of the country gives more importance again to cattle-rearing. In many parts where the Northern races have encroached on the domain of the Negro the ruling (Hamitic) race is devoted to cattle-rearing, and the subject Negroes to agriculture. The most widely-spread food-plants are the banana, cassava, and species of millet. The first is of special importance in the warm and moist region, which is unsuited to the growth of cereals, and the second is principally cultivated in the western savannah regions. Maize and rice are grown mostly in those parts which have a decided dry season at one portion of the year, and durra (sorghum or Guinea corn) is the favourite cereal in the lands bordering the desert on the south. In the north the principal cultivation is that of the date palm in the desert, and of the olive, fig, and vine in the coast strip north of the Atlas, which in many respects resembles the south of Europe.

The mode of life of the races of small stature scattered over the southern half of the continent differs from both of the above main types. These tribes are of nomadic habits, wandering about in search of game, which they kill with poisoned arrows.

Political and Social Characters.—In no continent is the tendency to the formation of states weaker than it is in Africa. This is in great measure a result of the general external uniformity of the continent, owing to which there are few sharply-defined subdivisions such as the peninsulas of Europe and Asia, or those formed by well-marked mountain ranges like the Himalayas. In the case of the Negroes this cause is supplemented by the natural characteristics of the race, which shows, perhaps, less aptitude for political combination than any other people. The Northern races, who undoubtedly stand on a higher level of culture than the Negro, show a much greater disposition to the formation of states, which are found on both sides of the desert zone, where the possibility of cultivation gives a greater fixity to the population. In the desert itself the pastoral life of the inhabitants, with the nomadic habits which it induces, tends rather to the formation of numerous family groups or clans under patriarchal rule, these being occasionally combined over a large area to form a sort of political confederation. Longstanding feuds are maintained between rival clans, and the inhabitants of the poorer regions live in great measure by plundering their more fortunate neighbours. In the southern half of Africa there hardly exists a really important kingdom, and where such has been formed it has never been of long duration. Instead of powerful kings,

FIG. 7.—WALL AND DITCH OF WEST AFRICAN TOWN.
(Photograph by Rev. J. T. F. Halligey.)

we find only innumerable petty chiefs, whose authority, in many cases, is restricted to a single village; a general state of mutual distrust is the result, and feuds are perpetually being carried on between neighbouring tribes or villages. The villages are generally surrounded with a mud wall, palisade, or thorn hedge with a single narrow entrance, often so low that it is necessary to stoop to enter. This state of things is one of the greatest obstacles to progress in civilisation. It has been largely encouraged by the slave trade, which has been the scourge of Africa for centuries.

Other characteristics of the Negro are a childlike nature, with great buoyancy of spirits, but little self-control or forethought; cruelty and superstition, resulting in human sacrifices, trials for witchcraft, etc. Affection between parents and children is strong, but the prevalence of polygamy, especially among the chiefs, leads to endless discords and other evils, such as raids for the seizure of women, on whom often falls almost the whole of the manual labour, both in the house and in the fields. Cannibalism is still prevalent in certain parts, having mostly originated in the difficulty in procuring means of subsistence. It is, however, found among tribes who in other ways occupy a fairly high position in respect of culture, and in whose case such an explanation does not hold. The Bantu Negroes, as a whole, may be placed on a higher level than those of the Sudan, and show a greater capacity for improvement.

Religion.—The Northern races, with hardly an exception, profess the religion of Mohammed, and an excess of religious fanaticism is to be seen among some tribes. The religious ideas of the Negroes are very primitive, consisting chiefly in the reverence for ancestors, and —especially in West Africa—for certain inanimate objects (fetishes), which are supposed to be in some way connected with spirits, over whom their possession gives a certain amount of influence. Where they have come in contact with the Northern races, they have to a large extent become converted, nominally, to Islam, which has rapidly extended its influence southwards of late years; little or nothing is, however, understood of the doctrines of that religion by the so-called converts. Much more has been done towards elevating the Negroes by the missionaries of the various Christian communions, who have of late years met with considerable success in their work.

Trade and Industries.—For many centuries trade has been carried on in North Africa across the desert for the interchange of the products of the countries to the north and south of that region, and the export of its two main products—dates and salt. Goods are carried by long strings of camels, known as caravans, by the help of which alone the desert can be crossed. In the Negro domain commerce has been in a very backward state until recent years, although it is supposed by some that ivory has been exported from Central Africa to the east coast since early times. Means of inter-communication have been almost entirely wanting, as there are no roads, only narrow winding footpaths connecting the villages, and forming in parts frequented routes for longer distances, along which, however, goods can only be transported on men's heads.

Industries are few, and hardly supply any articles of commerce. In Northern Africa goat's hair is woven into cloth, both for clothing and tents. In the southern half of the continent iron is worked wherever present, and iron implements often form objects of barter. The Negro, however, is not inventive, though he readily adopts methods and appliances which he sees others use. He shows a good deal of skill in the simpler handicrafts, such as basket-work, weaving, and the fabrication of wooden utensils.

The houses of the Negroes are of a very unsubstantial character, the walls being mostly constructed of upright posts interwoven with branches of trees, palm leaves, or grass, with only one opening to the outer air. They are usually circular, with conical roofs, sometimes so high as to be likened to extinguishers, but more often of a "bee-hive" shape. Rectangular huts are also found, the most peculiar form being the flat-roofed "tembe" of East Africa. In North Africa, and those parts of the Negro habitat within the influence of the Northern races, stone or clay dwellings are in use, and there the towns are surrounded by high and thick clay walls. The Negroes everywhere are clever at the construction of foot-bridges over the streams, often in the form of suspension bridges supported by creepers.

Population.—It is almost impossible to form any correct idea of the total number of the inhabitants of Africa, and the estimates given differ enormously. Some recent writers consider that the number is about 200,000,000. Even in the better-known parts there is great uncertainty, and the numbers given for Central Africa are little more than guesses. Along the rivers or in districts favourable for agriculture a fairly thick population has often been found, but these are probably separated by wide extents of thinly-peopled country, especially in the forest regions or arid steppes. The desert tracts are, of course, very sparsely peopled.

Recapitulation

We may conclude this chapter by summing up the broad characteristics of the principal natural regions into which the continent may be divided in accordance with the facts already mentioned. Beginning from the north, we have first a strip bordering the Mediterranean (mountainous to the west, more level and lower to the east) in which the conditions of climate, plant and animal life are, on the whole, similar to those of South Europe. To this follows the great desert of North Africa, a low plateau traversed in various directions by ranges of mountains, subject to intense heat, its barrenness only relieved by scattered oases, whose principal production is the date palm. It is peopled by Arab and Hamitic tribes following mainly a nomadic pastoral life. On the southern border of the desert the country begins to improve, and we come to a more or less fertile zone, still forming a comparatively low plateau, and in general composed of open country which admits the practice of both agriculture and cattle-rearing. In this zone, called the Sudan or land of the blacks, the Negro race begins to appear, but still with an admixture of Hamitic blood. These mixed peoples generally have the chief power in their hands, and are sometimes keen traders. To the east the valley of the Nile, which passes through these three zones in succession, forms a link between them, and in many ways constitutes a natural region in itself.

Farther south we come, in West Africa, to a region stretching along the shores of the Gulf of Guinea and spreading out over a wide area in the Congo basin, with a humid climate and no great variations of temperature, largely covered with forest, the special home of the oil palm and also of the anthropoid apes. Its inhabitants are all Negroes, with the exception of some hunting tribes of small stature, and, as a rule, practise agriculture. In the same latitudes to the east the most elevated region of Africa occurs, beginning with the highlands of Abyssinia east of the Nile valley, and widening southwards into an extensive plateau furrowed by the branches of a great north to south line of depression, parts of which, as well as some broader basins on the surface of the plateau, are occupied by large lakes. Isolated peaks or groups of mountains are scattered over the surface, mostly of volcanic origin. The prevailing form of landscape is that consisting of savannahs or steppes, roamed over by game and generally suited both for cattle-rearing and agriculture, the former practised in particular by Hamitic peoples, who have pressed southwards into the Negro domain. The peninsula (also inhabited by Hamitic tribes) which projects eastward from the East African plateau is generally lower and more arid, and is little suited for agriculture. The whole southern portion of the continent up to the mouth of the Zambesi and (farther west) about 12° S. has little to distinguish it from the East African plateau in point of altitude. Any large lakes that may have existed in former ages are, however, now dried up, while traces of volcanic action are rare. The amount of rainfall becomes less, and is confined to one season of the year; in the extreme south the climatic conditions are those of the temperate zone. The Bantu tribes of this region are more given to cattle-rearing than those farther north. In the South-West occurs a small arid region, similar to that of North Africa, occupied by the Hottentots and Bushmen, races of doubtful affinities, the former eminently herdsmen, the latter nomadic hunting tribes. Lastly, the great island of Madagascar forms a region quite distinct from the rest of the continent, the treatment of which, under all its aspects, may be reserved for a future chapter.

CHAPTER III

EXPLORATION

1. *General Considerations*

Relations with other Continents.—In the preceding chapter we have dealt principally with the native races of Africa, or those whose ancestors migrated into the continent so long ago that no historical account of their coming is preserved. We have now to see how the history of the continent has been modified within historic times by contact with peoples from without. In this connection the most striking fact is the great length of time during which Africa as a whole has remained sealed up, as it were, and hardly touched by outside influence. The history of North and South America presents a marked contrast with that of Africa in this respect. Although Africa contained one of the earliest centres of civilisation known (the Egyptian), and lay at the very doors of the most important empires of ancient and modern times, it remained almost unknown to the rest of the world, while America, and particularly South America, though only first made known to the inhabitants of the eastern hemisphere 400 years ago, became known in its broad outlines within half a century from its discovery. The strip of country lying along the northern coast of Africa must be excepted in this statement, for as in its physical features, so also in its history this region has much more in common with Europe and Western Asia than with the rest of Africa, the Mediterranean, by the facilities for navigation and trade which it affords, having promoted the mutual intercourse of all the countries bordering on its shores.

We see at once that the great obstacle to the extension of this intercourse beyond the narrow strip along the coast has been the great desert of the Sahara, which, before the introduction of the camel, formed an insuperable barrier to an advance southward, and even with its aid can only be crossed with difficulty and danger. The

Fig. 8.—Ruined Mosque, East Africa.
(Photograph by Sir John Kirk.)

nature of the inhabitants (itself largely a result of their environment) has co-operated with that of the country in bringing about this result. Another cause which acted in the same way was the extension of the Saracen power to North Africa in the seventh century A.D., the result of which was practically to close the whole of that region to Europeans, whilst the "Barbary Corsairs," who early in the sixteenth century established themselves in the principal ports of the coast, and were long a scourge to the merchants of the Mediterranean, proved a similar barrier.

With these powerful obstacles in the way of intercourse

with Africa from the north, it is not surprising to find that the relations with Southern Asia in early times were much closer. For a long period, in fact, Eastern Africa was closely connected with the Arab Power which had its centre in South-West Arabia, where Aden now stands, and which carried on an extensive trade with most of the countries bordering on the Indian Ocean. Evidence of such a connection is supplied by the remains of ancient buildings on the east coast, and of some still older in a gold-mining district of South-East Africa, which were certainly not the work of the native inhabitants. A similar political connection in later times was that between Zanzibar and the Sultans of Maskat, from the seventeenth to the middle of the nineteenth century, while in our own day trade has fallen in an increasing degree into the hands of natives of India, from amongst whom also the police forces of British Tropical Africa are largely recruited.

Isolation of Central Africa.—This, however, applies principally, if not entirely, to the coasts, and the fact remains that the whole centre of the continent had, until the nineteenth century, hardly any dealings with the outside world as known to history. The main causes of this isolation are connected with the general configuration of Africa. The simplicity of outline necessarily increases the inaccessibility of the interior parts, and to this is added the want of access from the sea by navigable rivers, for, as already noticed, almost all the African rivers are broken by cataracts on their passage through the ranges which bound the central highlands. The unhealthiness of the coast-lands has also stood in the way of penetration by natives of other continents, and, again, the political status of Africa has not been favourable to intercourse with outsiders. There have been in the interior no rich or powerful empires either to tempt adventurers like Cortez and Pizarro, or to afford support and protection to traders such as was enjoyed by Europeans on their first arrival in Southern Asia. In the complex subdivision of the country among petty chiefs, regarding each other with hatred and jealousy, it is enough to be on friendly terms with one ruler to incur the enmity of his next neighbour, while the generally savage and warlike character of the inhabitants has proved a much more formidable obstacle than was present in America.

Africa is, moreover, strikingly wanting in a ready supply of products capable of supporting extensive trade relations with other countries. The staple products of the continent have consisted from the earliest times of gold, ivory, and slaves, the first of which has never been methodically worked by the inhabitants themselves, and none of the three could ever be of the same importance as the spices, silks, and muslins of the Far East, or the vast supplies of precious metals and plantation products of America. Nor, on the other hand, has the culture of the inhabitants been high enough to create much demand for foreign goods in exchange.

2. *Ancient and Mediaeval Geography*

The Knowledge of the Ancients.—Our present knowledge of the interior of Africa has been almost entirely the work of the nineteenth century, and by far the greater part has been accomplished within the latter half of that period. It is by no means easy to determine how much of the continent was known at various times to the successive representatives of civilisation. The knowledge of the ancient Egyptians, the earliest race of whose civilisation we have any account, was probably confined (in addition to the northern coasts) to the northern half of the Nile valley and to the coasts of the Red Sea, and perhaps part of the shores of the Indian Ocean beyond. That of the Phoenicians and Carthaginians, both essentially maritime nations, was probably more extended, at least with regard to the coasts, and the story mentioned by Herodotus that a Phoenician vessel circumnavigated Africa, starting from the Red Sea and returning by the Mediterranean, may quite possibly be true. The Carthaginians possessed settlements not only along the Mediterranean coast, but also outside the Straits of Gibraltar, and their boldest voyager, Hanno, is supposed to have reached as far south as some point on the coast of Upper Guinea. The knowledge acquired by these peoples found its way to the Greeks, who also themselves formed settlements in North Africa, and the geographers of that nation and the Romans, from Eratosthenes onwards, show in their writings a considerable knowledge of the various branches of the Nile, besides mentioning races of small stature, such as have been brought to light within recent years. During the Roman epoch Greek and Roman travellers ascended the Nile, and an expedition sent by Nero seems to have reached a point very far to the southward. The knowledge existent at that time is summarised in the geography of Ptolemy, who flourished at Alexandria in the middle of the second century A.D., and who seems to have heard something of the country around the Nile lakes, possibly derived from Arab and Greek merchants, who traded on the east coast in those days.

Arab Epoch.—That further information did not reach Europe from this quarter may be ascribed to the rise of the Saracen Empire, which formed a barrier between Europe and the east. The extension of Arab influence over the greater part of Northern Africa (while a hindrance to European intercourse with the central regions) enlarged the bounds of knowledge by the promotion of trade with the countries south of the Sahara, on which the writings of Arab historians and geographers from the tenth to the sixteenth century supplied the first definite information. During this period Mohammedanism took root in the countries of the Western Sudan, and has ever since been gradually extending southwards. On the east coast the Persians as well as the Arabs formed settlements, but neither of these peoples seem to have penetrated far into the interior of the southern part of Africa.

3. *The Beginning of Modern Discovery*

The Portuguese Voyages.—The credit for the initiation of modern discovery belongs to Prince Henry of Portugal (surnamed "the Navigator"), grandson, on the mother's side, of the English John of Gaunt. His connection with Africa (like that of the Portuguese nation) began with the siege of Ceuta in 1415, and from this time onwards he devoted himself to the discovery of a sea-route round Africa to the East, the treasures of which might by this means be shared by his country. With admirable method and steadfastness of purpose he despatched expedition after expedition, each extending a little the bounds of knowledge on the west coast of the continent, until at his death, in 1460, Portuguese ships had almost reached the point where the coast of Guinea bends eastwards, while trade had already been opened with the countries visited. It is difficult for us to understand the feelings of mystery and fear with which unknown regions were viewed in those days, and great courage was required in the old sailors who undertook these voyages. It is wonderful that so much could be done in the small and frail vessels then in use, in which the actual danger incurred was much greater than in the ships of the present day, so that the merit of these achievements is not to be judged merely by the extent of country explored. After the Prince's death the work was carried on without interruption, and the Congo mouth was reached in 1482 by Diogo Cão, the Cape of Good Hope doubled by Bartholomew Diaz in 1483, and finally in 1487-88 the crowning success was attained by the voyage of Vasco da Gama to India. On the east coast he found flourishing Arab settlements at Mozambique, Mombasa, and Melindi, at the last of which he met with a most friendly reception, and obtained pilots from the King, whose guidance was of great use in the voyage across the intervening ocean to India. Vasco da Gama's successors discovered the great island of Madagascar, called by them Ilha de São Lourenço, and established forts and settlements along the whole east coast of Africa. On the west coast, too, the Portuguese influence had already been established in the kingdom of Congo, and other settlements soon followed. An account of Congo was published by F. Pigafetta in 1591, from the reports of Duarte Lopes, a Portuguese who had spent nine years in those parts.

Before the voyage to India by the sea-route, the vague

accounts then current of the empire of Prester John—a name originally applied in the twelfth century to a mythical Christian Emperor in Asia, but subsequently transferred to the King of Abyssinia—had induced the Portuguese to send an embassy to that country by way of the Red Sea. This was entrusted to Pedro de Covilham and Affonso de Payva. Covilham, after visiting India, made his way to Sofala, on the African coast, and afterwards succeeded in establishing friendly relations with the King of Abyssinia, in which country he resided until his death. This was the beginning of a long connection of the Portuguese with Abyssinia, during which missionaries collected much information as to the geography of the Abyssinian highlands. One of these was Jeronymo Lobo, who visited the country near the source of the Blue Nile early in the seventeenth century, and wrote an account, the translation of which into English was long afterwards the first literary venture of Dr. Samuel Johnson. Progress in this direction ceased, however, about the end of the seventeenth century, and was only resumed many years later.

Many attempts have been made to prove, from maps published in the sixteenth and seventeenth centuries, which show a network of interlacing streams and various lakes giving rise to the Nile, Congo, and Zambesi, that the interior of Africa was well known to the Portuguese in the early days of their colonies; but though it is very probable that a certain amount of knowledge gained was subsequently forgotten in the absence of published accounts by persons capable of accurately describing what they had seen, the arguments brought forward to prove a greater knowledge have so far been inconclusive. It was not until the end of the eighteenth century that the first scientific expedition into the interior was despatched by the Portuguese.

Work of other Nations.—Meanwhile other European nations, the British, French, Dutch, and Danes, had begun to frequent the coasts, especially those of Guinea, in quest of gold and slaves. In the country on the Senegal and Gambia many attempts were made to open a way into the interior, where the city of Timbuktu, of the riches of which the Arab writers had said so much, was the goal aimed at. The belief that the Niger flowed westward to the sea as the Senegal was another reason why this part of the coast was preferred to others. In particular, the Frenchmen, André Brue (1697-1725) and Compagnon (1716), penetrated into the interior of Senegal, while Messrs. Thompson and Jobson ascended the Gambia under the orders of an English Company.

Although the Portuguese fleets in their voyages to India often put in to Table Bay (then called Saldania Bay, after Antonio de Saldania, who discovered it in 1507) for the purpose of watering, the wealth of India was so much greater an attraction that they established no permanent settlements in South Africa. The English and Dutch, who, from the end of the sixteenth century, became the rivals of the Portuguese for the trade of the East, likewise often touched there, but it was not until 1652 that a permanent occupation of the site of Cape Town was effected by the Dutch. Gradually, and at first in spite of positive discouragement by the Dutch authorities, this settlement expanded into a small agricultural colony, which was reinforced by French Huguenots and other Protestant refugees after the revocation of the Edict of Nantes. As the bounds of the settled districts were enlarged, regular surveys of the country were executed, and some journeys were made into the interior by private individuals such as Peter Kolbe (1705-13) and Le Vaillant (1783-85).

4. *The Exploration of the Interior*

Bruce's Journey. The earliest great journey of exploration, which, though not immediately followed up by others, may be said to have heralded the advent of an increased interest in the geography of Africa, was that of James Bruce, who between 1768 and 1772 made his way from Massaua to the source of the Blue Nile, in which he imagined he had found the principal source of the Nile

of Egypt. Bruce's journey was the subject of much doubt
and ridicule at the time, and it was only after many years
that the general truthfulness of his account was acknowledged.

The Niger Problem.—The event, however, by which
the modern exploration of Africa was definitely ushered in
was the foundation in England, in 1788, of the African
Association, largely due to the initiative of Sir Joseph
Banks, the companion of Captain Cook on his first voyage.
Although nearly three centuries had elapsed since the coasts
of Africa had been made known to Europe, and in spite
of the trading relations established on them by various
European nations, the known part of the continent was
limited (but for a slight advance in Abyssinia and Senegambia) to the merest fringe round the margin, the whole
interior remaining a blank on maps constructed from
definite information. To remedy this state of affairs was
the object of the newly-formed Association, which chose as
its first task the problem of the course and termination of
the Niger. This river, though mentioned under that name
by Ptolemy and other early writers (and possibly even
referred to by Herodotus), had been the subject of unending
speculations, which in turn made it flow eastwards to the
Nile, or westwards to the Senegal, or supposed it to end
in a lake in the centre of the continent. The first decisive
result was obtained by Mungo Park, who (1795-97) reached
the Upper Niger from the Senegal, and proved its easterly
direction by actual observation. On a second journey
(1805) he descended the stream yet farther into the
interior, but his canoe running on a rock in the presence
of hostile natives at Bussa on the middle course of the
river, he and four European companions were drowned,
and the problem as to the termination of the Niger
remained unsolved. In 1816 a double attempt was made
to decide the question both from the Upper Niger and
the Congo (into which it had been suggested that the
river flowed) by Major Peddie and Captain Tuckey
respectively, but both journeys proved complete failures.

Meanwhile attempts had also been made to penetrate
the continent from the north. Mr. W. G. Browne made

an important journey from Egypt to Darfur in 1793-96. In 1798 Hornemann, an agent of the Association, journeyed from Cairo to Murzuk, and in 1800 started from Tripoli to cross the Sahara, but died without sending home any further news. In 1819 Messrs. Ritchie and Lyon followed in his footsteps as far as about 24° N.L. In 1822-24 the great expedition of Denham, Oudney, and Clapperton crossed the Sahara and discovered Lake Chad and the mouth of its main feeder, the Shari. In 1825 Major Laing succeeded in reaching Timbuktu from the north, being the first European of whose visit anything definite is known, but he was shortly afterwards murdered. Three years later the Frenchman Réné Caillié reached the same town from the south-west, and successfully crossed the desert and the Atlas range to Tangiers. These journeys paved the way for the final solution of the Niger problem. Captain Clapperton, who had accompanied Major Denham in 1822, went back to the Sudan in 1825-27, starting from the Gulf of Guinea, and though he died without actually tracing the river to the sea, his servant, Richard Lander, returned in 1830 and verified its connection with the streams on the Benin coast, suggested as early as 1802 by a German savant named Reichard. The mouth of the Niger could not be found from the sea because it is broken up into many small branches (forming the delta), which run through dense swampy forests.

The solution of this problem, and the exploration of portions of the Sahara and Sudan, was the most important work accomplished from the founding of the African Association until it was merged with the newly-formed Royal Geographical Society in 1831.

Other Explorations before 1850.— Knowledge was gradually extended in other directions also. The first scientific expedition sent by the Portuguese into the interior was that under Dr. J. de Lacerda in 1798, which threw a new light on the countries far inland from the colony of Mozambique. The first known communication between the colonies on the east and west coasts was effected about this time by two native agents of the authorities in Angola, generally known as the " Pombeiros " (bondsmen).

The knowledge so gained was, however, allowed to slumber for years. In 1805 and 1809-10 Henry Salt explored Abyssinia, which had not been visited since Bruce's time. With the extension of Egyptian influence southwards under Mehemet Ali, travellers in the Nile region became more numerous. The principal journey was that of Caillaud and Letorzek (1821-22). The southern half of the Abyssinian highlands was explored between 1832 and 1848 by a variety of travellers, including the missionaries Gobat and Krapf, Dr. Beke (1840-43), and the brothers D'Abbadie, who penetrated southwards to Kaffa in the Galla countries. On the White Nile a great advance was made by the expeditions sent by Mehemet Ali in 1840-41, the second of which reached 4°42′ N., and shortly after this traders began to establish themselves in these parts, while in Europe geographers began once more to discuss the question of the sources of the river. In Algeria the French occupation in 1830 paved the way for future progress from this starting-point. In South Africa the region north of the Orange River began to be visited within a few years of the British acquisition of the Cape (1795), Bechuanaland being first reached between 1801 and 1806, and the Upper Limpopo in 1812. The "Trek" of the Boers in 1835 greatly helped to extend the bounds of knowledge, and about the same time Dr. Andrew Smith, an accomplished zoologist, penetrated as far as the Upper Limpopo. Captain J. E. Alexander traversed the great Namaqua and Damara countries near the west coast in 1837.

About the middle of the nineteenth century, therefore, some advance had been made from the south, much more, however, in the north from three principal starting-points, viz. Senegambia, Tripoli, and Egypt with the Red Sea coasts. From both east and west in Equatorial Africa hardly any advance had been made. The years subsequent to 1850 were to see not only an increased advance both from north and south, but the commencement of penetration from the east, a route which soon rivalled all the rest in importance. The west did not serve as a starting-point for exploration until some time later.

Renewed Efforts.—In 1849 signs of increased activity

manifested themselves in three separate quarters. James Richardson, who had already made a successful journey in the Northern Sahara, laid before the British Government his plans for a great expedition across the desert; Dr. Livingstone, who since 1840 had laboured under Dr. Moffat as a missionary among the Bechuanas, made his first exploring journey northwards, discovering Lake Ngami;

Fig. 9.—Timbuktu. (After Barth.)

and on the east coast Krapf and Rebmann verified the existence of snowy mountains in the interior, and collected information as to a reported inland sea. As the northern exploration had more points of contact with previous work than that in the south and east, it will be well to speak of it first.

North Africa. Mr. Richardson, accompanied by the two German volunteers, Drs. Barth and Overweg, set out from Tripoli in March 1850, and, crossing the desert from

Murzuk by the westerly route *via* Air, instead of that *via* Bilma followed by Denham and Clapperton, had almost reached the fertile countries of the Sudan when death overtook him. Dr. Barth hereupon took the command, and during the next five years traversed the Central Sudan in all directions, studying the physical, political, and social aspects of the country, which on his return he described in a work hardly to be matched in African literature for the extent and variety of its information. During this period he reached from the north the Benue, the great eastern tributary of the Niger (shortly after ascended from the sea by Dr. Baikie); crossed the Shari and visited the capital of Baghirmi; and finally made his way to Timbuktu, returning along the Middle Niger, then first laid down on the map with any degree of accuracy. This region was not visited by any other traveller until forty years had elapsed. Dr. Barth was followed across the Sahara by Dr. Vogel (1854), who died in attempting to penetrate eastwards through Wadai, while the northern parts were explored in 1859-61 by Duveyrier. In 1866 Gerhard Rohlfs, who had already made himself known by his journeys south of Marocco, crossed the desert by the Bilma route, and continued his way through the Sudan to the Guinea coast at Lagos. He was followed in 1869 by Dr. Nachtigal, who, besides exploring the Tibesti highlands and a depression north of Lake Chad, threw much light on the countries south of the lake, returning by an easterly route through Wadai, and thus supplying the first connection between the Central Sudan and the Nile basin. Numerous French travellers have since explored the regions south of Algeria, but the next journey across the desert was that of Dr. O. Lenz, who in 1880 reached Timbuktu from Marocco by a route across the little-known Western Sahara.

South and East Africa.—The success of his first effort led Dr. Livingstone to undertake more extensive journeys. In 1851 he discovered the Zambesi flowing in the centre of the continent, though its identity was not absolutely proved until a few years later. In 1853 he ascended the stream almost to its source, reached the

Portuguese colony of Angola on the west coast, and, returning, traced the Zambesi through the greater part of its course to the sea, being thus the first to cross the continent from west to east. In his next journey (begun in 1858) he filled up the gaps in his survey of the river, and first shed clear light on Lake Nyasa, the existence of which had been known previously from Portuguese traders. It was reached very shortly afterwards from the east by the German traveller Dr. Roscher. Subsequent explorers in South Africa may be here mentioned before proceeding to more northerly regions. In the western parts Galton and Andersson opened the way from Walfish Bay into the interior (1850-59). More in the centre Chapman and Baines were among the earlier pioneers, and farther east, in Matabililand and the Matoppo Mountains, Moffat and Karl Mauch, the latter of whom first discovered gold in those parts. Erskine in 1868 traced the Limpopo to the sea, and subsequently explored the region to the north, while Mr. Selous and Dr. Holub made extensive additions to the knowledge of the central Zambesi basin.

On the east coast the reports of the missionaries led to the despatch by the Royal Geographical Society, in 1858, of an expedition under Captains Burton and Speke, which discovered Lake Tanganyika in the centre of the continent. On the return march Captain Speke paid a flying visit to the northern Lake of Ukerewe, named by him Victoria Nyanza, which he at once pronounced to be the principal source of the Nile. To verify this assertion he was again despatched to East Africa, this time with Captain Grant, and made his way round the western side of the lake to the native kingdom of Uganda on its northern shore, where he discovered an outlet flowing northwards in the direction of the Nile. Meanwhile Mr. (afterwards Sir Samuel) Baker, following in the steps of John Petherick and other traders who, like the Arab slave-raiders, had been pushing southwards along the Upper Nile, had advanced to meet the travellers from the north, and hearing that a second Nile lake remained unvisited, continued his journey southwards, and was rewarded by the discovery of the Albert Nyanza. Thus the main facts of the hydrography of the

Nile were at last known, after over 2000 years of inquiry and conjecture.

The Congo Basin.—While the eastern half of Equatorial Africa had thus within a few years become known in its broad outlines, the west still remained almost a blank on the map. To remedy this state of things Dr. Livingstone again set out in 1865 to clear up the still doubtful points in the hydrography of South Central Africa, especially with regard to the exact part played by Lake Tanganyika, and the southward extension of the Nile basin. Starting from a point near the mouth of the Rovuma, and passing round the south end of Lake Nyasa, he struck north-west and came upon a system of rivers and lakes with a general northward flow, which finally formed a vast river called Lualaba. His farthest point on this stream was the Arab mart of Nyangwe, lately founded by the slave-hunters, who were then beginning to extend their operations against the defenceless natives of the countries west of Lake Tanganyika. Northward of this was an exceedingly difficult forest region, into which he was unable to penetrate, and death finally overtook him before he had been able to clear up the mystery of the ultimate destination of the river, which he had at first held to be the Nile. Latterly doubts had sprung up in his mind whether he had not struck the head-waters of the Congo, the great size of which had long before been remarked on but not widely recognised. This view was adopted by Lieutenant Cameron, who, chosen to carry aid to Livingstone, proceeded with his journey in the interests of geography on hearing of the death of the great explorer, and after visiting Nyangwe traversed the unknown region between that point and the west coast in a south-westerly direction, and thus, for the first time in history, effected the crossing of the continent from east to west. During the first part of this journey he had circumnavigated the southern half of Lake Tanganyika, and discovered on its west shore a break in the surrounding wall of mountains, through which he maintained that the surplus water found its way to the Lualaba.

A little before this (1868-71) an important journey had been made from the north by Dr. G. Schweinfurth, who crossed the south-western limit of the Nile basin and came upon a large river (the Welle) flowing westwards, the termination of which long remained a problem, but which ultimately proved to be a tributary of the Congo.

During Dr. Livingstone's long and arduous explorations he had for some years been lost sight of in the interior until, in 1871, Mr. H. M. Stanley—sent by Mr. J. Gordon Bennett of the *New York Herald*—found him at Ujiji, and replenished his exhausted supplies. In 1875 Mr. Stanley was commissioned by the proprietors of the *Daily Telegraph* and *New York Herald* to attempt to solve the remaining problems of Central African geography. Reaching the Victoria Nyanza by a new route, he circumnavigated the lake, and thus disproved the theories of those who had thought it was a group of small lakes; he then struck westwards and discovered the arm of a new lake south of the Albert Nyanza, of which he at the time believed it to be a part; then, after a re-examination of Lake Tanganyika and its outlet, he too reached Nyangwe, and journeying through trackless forests, past cataracts and warlike cannibal tribes, succeeded in tracing the Lualaba round its great northern bend until it brought him to the known part of the Congo close to the Atlantic Ocean. By this great journey across the last really important unknown area on the surface of the globe (excepting only the uninhabited polar regions) he had made known sixteen hundred miles of the course of the greatest African river, and disclosed a fluvial system surpassed only by that of the Amazon in South America.

Recent Activity.—The remaining history of African exploration is concerned rather with the filling in of details than with the record of startling discoveries. A general interest had now been aroused among European nations, which was further stimulated by the formation of an International Association for the opening up of the continent under the auspices of the King of the Belgians, the agents of which commenced operations both on the east and west coasts. The French (who even before

Stanley's journey had been exploring the interior of their colony of the (Gabun) now pushed on into the Congo basin, under S. de Brazza and others. Germans and Portuguese penetrated into the interior from the west coast, somewhat more to the south, the former represented by Drs. Pogge and Buchner, Lieutenant Wissmann and others, and the latter by Serpa Pinto and Messrs. Capello and Ivens. Two British expeditions were brought to a successful issue in East Central Africa by Joseph Thomson (who succeeded to the command of the first on the death of Mr. Keith Johnston), one to the head of Lake Nyasa and thence to Tanganyika, the other through the land of the dreaded Masai to the Victoria Nyanza, by way of the great East African trough. The interior of Mozambique was explored by Mr. H. E. O'Neill, British Consul at that port, and Somaliland was penetrated for the first time by Mr. F. L. James. The Italians Cecchi, Chiarini, and others explored Abyssinia and the Galla countries, while Dr. Matteucci and Lieutenant Massari crossed over from Suakin to the mouth of the Niger (1880-81).

Exploration has of late years been closely connected with the extension of political influence over the continent, which will be described in the next chapter. Among the host of explorers of all nations who have vied with each other in extending the bounds of knowledge, a few only can be mentioned whose journeys have led over the largest amount of unexplored country. Such are, in the Congo basin, Wissmann, Grenfell, Von François, Delcommune, Vangele, and many others; on the watershed between the Nile and Congo, Dr. Junker and Mr. Stanley (the latter of whom traversed the great equatorial forest and discovered the great snowy mountain Ruwenzori on his journey for the relief of Emin Pasha); between the Congo and the Niger and Lake Chad basins, the French explorers J. de Brazza, Mizon, Crampel, and Maistre; in the Niger basin, the Frenchmen Binger and Monteil, and quite recently Decœur and Toutée, as well as the Germans Grüner and Passarge, and the English Captain Lugard; in the Galla and Somali countries, the Italians Robecchi, Bòttego, and Prince Ruspoli, and more recently still, the American traveller, Dr. Donaldson Smith; in East Africa, the Austrians Teleki and Von Höhnel (the discoverers of Lake Rudolf), and Dr. Gregory, in Masailand and neighbouring countries; and on the southern border of the same, Dr. O. Baumann (the discoverer of Lakes Manyara and Eyassi). Lastly, in the most central region, north of Lake Tanganyika and westwards to the Upper Congo, Count von Götzen has lately opened a new route from east to west, and discovered Lake Kivu and the smoking volcano, Mount Kirunga.

As a result of all this exploring energy, there are now few tracts of any extent not traversed by Europeans, except in the Sahara, where vast regions of desert still remain *terrae incognitae*, and will probably be so for many years to come. Much is still left to be done, however, in the way of filling in of minor details and the execution of more accurate surveys, for much of the mapping of the interior still depends solely on rough route-surveys with the compass. The extension of a more scientific system of mapping to this continent (especially needed for the definition of international boundaries) must soon take the place of great exploring journeys.

CHAPTER IV

POLITICAL RELATIONS

1. *Native States*

A RESULT of the slight capacity for founding strong or stable kingdoms possessed by the races of Africa, to which allusion was made in the second chapter, has been that any parts of the continent which have been brought into permanent relations with the rest of the world have, as a rule, fallen under the dominion of foreign races more gifted in this respect. Thus, after the decline of the ancient Egyptian civilisation, which forms the only important exception to this rule, North Africa fell successively into the hands of Phoenicians, Greeks, Romans (of whose occupation countless traces are still to be seen in Algeria and Tripoli), and Saracens, all of them hailing from neighbouring regions of South-West Asia or Europe. The Saracens established themselves so firmly in North Africa between the seventh and eleventh centuries A.D., imposing the religion of Mohammed on the former inhabitants, that they soon ceased to be foreigners, and their presence has been the ruling influence almost to the present day, even where the actual sovereignty has fallen into other hands. The east coast, too, has from very early times fallen under the control of the inhabitants of Southern Arabia, and to the Arab influence introduced in these two quarters may be ascribed the foundation of almost every state which has maintained its importance since the advent of the nations of modern Europe.

North Coast.—The Arab rule on the north coast gave

place in Egypt to that of the Mamelukes in the thirteenth century, and in Algeria and Tunis alternated with that of native Berber dynasties until the sixteenth, when all these countries fell under the dominion of the Turks, and have since remained (in name at least) under foreign rule. Marocco, on the other hand, has maintained its independence under Arab or Berber dynasties down to the present day, and is the only true native state now remaining on the north coast.

Sudan.—Besides establishing themselves in the north of the continent, the Arabs crossed the Sahara during the centuries subsequent to their first coming, and extending the Mohammedan religion over the Western and Central Sudan, supplied the stimulus for the formation or re-invigoration of kingdoms, which flourished for a time and then gave place to others. Most of these kingdoms arose among Negro populations, who became converted to Islam. Such were those of Songhai, on the Middle Niger, with Gogo and Timbuktu as its principal centres, which embraced that religion about 1000 A.D., and Melli, on the Upper Niger, among the Mandingos, which took the place of the old pagan kingdom of Ghana. The kingdom of Bornu, west of Lake Chad, which has held its own until recent times, was founded among the Kanuri, apparently a mixed race. Other states which came into prominence at a later period were those of Darfur, Wadai, Baghirmi, etc., farther to the east. These remained longer under pagan rulers, but Mohammedanism finally prevailed, and the states acquired a new importance in the sixteenth and seventeenth centuries. Quite recently they have been subject to vicissitudes which will be mentioned in a later chapter. In the Western Sudan the Fulahs (also called Fulbe, Fellata, or Fellani, probably a Hamitic people which has adopted a Negro language), who had till then remained a subject race of herdsmen, began a career of conquest at the beginning of the nineteenth century, and under their religious teacher, Othman, overran the whole region from the Upper Niger to Bornu. The empire of Sokoto, which they founded, did not long remain united, and only the state of Gando, on the Middle Niger, has remained in semi-dependence on the kingdom of Sokoto proper, between the Niger and Bornu.

East Africa.—In the north-east of the continent we find a native state, that of Abyssinia, which in like manner shows the effect of Semitic influence. This, as was natural from the greater proximity to Arabia, began to be felt at a much earlier date than in North Africa, centuries, in fact, before the rise of Mohammedanism. The Himyarites, who inhabited the south-west of Arabia, founded the port of Adulis, on the African coast (a little south of the modern Massaua), at an early date, and soon established themselves on the Abyssinian highlands, founding as the centres of their power one city after another, the ruins of which are still to be seen in the present province of Tigré. The best-known of these was Axum, which owed much of its importance to Greek influences, and through which an important trade-route led from Adulis into the interior. The introduction, in the fourth century, of Christianity (which has survived in a debased form until the present day) led in time to antagonism with the Mohammedans, and consequent loss of power. The Portuguese activity in the fifteenth and sixteenth centuries has already been referred to. In more recent times the Gallas and other Hamitic tribes have encroached from the south, establishing themselves in Shoa, which has lately given a ruler to Abyssinia in the person of Menelek, the present Emperor.

The early Arab settlements on the east coast fell into the hands of the Portuguese soon after their first arrival, but in the seventeenth century the Imam of Maskat made himself master of Zanzibar, which about the middle of the nineteenth century became virtually an independent state, the only one on the east of the continent south of Cape Guardafui which has been of any importance in recent years.

Negro States.—The purely indigenous states of Africa have been very limited in numbers. The ancient pagan kingdoms in the Sudan have been already alluded to. In later times reports—doubtless much exaggerated—reached the Portuguese of a powerful emperor called Monomotapa in the region south of the Zambesi, and also of a region called Monemoezi or Monoemugi (by some supposed to represent the modern Unyamwezi) north of that river. Almost the only kingdoms deserving the name that have existed in our own times are the sanguinary despotisms of Ashanti and Dahome on the Guinea coast, and the domains of a few chiefs among the Bantu, more important than ordinary, such as the Muata Yanvo, in the interior of Angola; Kazembe, on the Upper Congo; Lo Bengula, in Matabililand; Chaka and his successors, in Zululand; and the kings of Uganda and the Monbuttu farther north. The power of such chiefs, though considerable for a time, is seldom of long duration.

The native republic of Liberia, on the west coast, holds a unique position among African states. It began its existence under the auspices of an American Society as a settlement of freed slaves, who, it was hoped, might help to civilise the natives of the neighbouring regions—a hope that has hardly been realised.

2. *European Possessions before 1880*

The Tropical Coasts.—The earlier territorial acquisitions of European nations after the discovery of the coasts by the Portuguese were touched upon in the previous chapter, and a brief recapitulation will suffice here. The Portuguese dominions soon included—in addition to isolated posts at Arguin Bay, north of the Senegal, San Jorge da Mina, on the Gold Coast, and other points in Upper Guinea—almost the whole coast between the Congo and the Kunene on the west, and from Sofala to Cape Guardafui on the east, extending some distance inland in Congo and Angola and on the Lower Zambesi. The first named posts were lost soon, and the northern half of the East African possessions, which were never very firmly held, in the seventeenth century. In spite of the efforts of Portugal to monopolise the continent, other nations entered the field after a time, and English, Dutch, French, Danish, and Prussian merchants visited the Guinea coasts in quest of gold and slaves, trading companies being formed for this end and forts established, including some wrested from the Portuguese. Many of these changed hands more than once. England had posts on the Gambia and on the Gold and Slave Coasts, but Lagos, on the latter, which has lately proved one of the most flourishing British settlements in West Africa, was not acquired till 1861. The Danes and Dutch also had possessions on the Gold Coast, which were acquired by Great Britain in 1850 and 1871 respectively. The French devoted their energies principally to the countries bordering on the Senegal, where they pushed their posts some distance into the interior. They also made repeated but unsuccessful attempts to establish themselves on the east coast of Madagascar. Much later, in 1842, they gained a footing on the Gabun, which was destined to lead to important results in course of time. The possessions of Spain—whose sphere of action had been fixed in the other hemisphere by the Bull of Pope Alexander VI.—were long confined to the Canary Islands (conquered in the fifteenth century). The island of Fernando Po, in the Gulf of Guinea, was, however, ceded to her by Portugal in 1778.

In **South Africa** the Dutch founded a station on the site of Cape Town in 1652, around which an agricultural colony in time sprang up, but this finally fell into the hands of the British in 1806 (after a temporary occupation from 1795 to 1802), and a part of the Dutch Boers subsequently "trekked" north and founded new states outside the British jurisdiction, which was afterwards extended to the colony of Natal, on the south-east coast.

The **North African** States (except Marocco) became, as already mentioned, subject to external control—that of the Turks—in the sixteenth century, though still retaining rulers of their own. In 1558 the Turks drove the Portuguese out of the Red Sea, and laid claim to the whole of the coasts, over which they (and subsequently Egypt, as the representative of Turkey) exercised a certain amount of control until recent years. In 1830 Algeria was conquered by the French, and Tunis fell under French influence in 1881. Egypt acquired a partial independence of the Turks in the first half of the nineteenth century under Mehemet Ali, who extended his sway southwards to Khartum, at the junction of the Blue and White Niles. The boundaries of Egyptian dominions were still further extended under his grandson Ismail, as far as the Albert Nyanza, through the instrumentality of Sir S. Baker and General Gordon. In 1882 a military revolt under Arabi Pasha led to the intervention of Great Britain, which has since maintained a military occupation of the country and exercised a control over the Government. The southern provinces have, however, been lost since 1884, owing to the revolt of the fanatical leader termed the Mahdi.

Position in 1880.—The above-mentioned acquisitions were practically all that had been made by European powers down to about the year 1880, from which date the commencement of a new era may be reckoned. Explorers (mostly British) had for some years been traversing the interior regions, but, except in the case of Egypt and the Sudan, little idea had been entertained of acquiring political influence in the regions visited, the majority of explorers having started from coasts already subject either to native rulers, such as the Sultan of Zanzibar, or to the Portuguese, who had themselves long remained inactive. The French alone were advancing from bases of their own in their possessions on the Senegal and Gabun. In the Nyasa country private enterprise, represented both by Missionary Associations and the African Lakes Company, founded for purposes of trade in 1878, continuing the work begun by Livingstone, had widely extended British influence, but as yet no definite political status had been acquired.

3. *The Modern Partition of Africa*

Beginning of a new Era. The interest in Africa aroused by Dr. Livingstone's labours and heroic death had, in 1876, induced King Leopold of Belgium to summon a conference in Brussels, which resulted in the formation of an International Association for the opening up of the continent, and almost at the same time a stimulus to increased activity was supplied by Stanley's

great journey down the Congo. In Germany also the
formation of an African Association was the outcome of
a general desire for new outlets for the growing trade
and emigration of that country. This soon led to more
definite action in the way of territorial acquisitions, the
first of which was the strip of land on the south-west
coast, north of the Orange River. The entry into the
field of this new competitor, eager to lay hold of any
ownerless tracts on the African sea-board, made other
nations bestir themselves to secure regions where, though
their influence was paramount, no definite annexation of
territory had been made. Thus in 1884, Germany having
already appropriated the district of the Cameroons, in the
Gulf of Guinea, where British traders had long been settled,
Great Britain hastened to secure control of the Lower
Niger, where, after a period of rivalry with the French,
a British Commercial Association (the "National African
Company") had finally obtained the monopoly of trade.

Meanwhile the activity of the International African
Association under Mr. Stanley on the Congo, towards
which the French were also pressing from the Gabun, and
to which Portugal laid claim by virtue of her ancient
relations with the kingdom of Congo, directed the attention of Europe to this region, and towards the end of 1884
an International Conference met at Berlin to regulate the
steps which should be held necessary to secure political
control over African territory and other points connected
with European trade with that continent. In particular a
so-called "Free-trade area" was defined, embracing a zone
of the continent from sea to sea, and including the Congo
basin, with adjacent parts of those of the Nile and Zambesi.
Another event which resulted in some way from the Conference was the formation of the Free State of the Congo,
the outcome of the International Association, the sovereign
rights of which to almost the whole Congo basin were
recognised at the time by all the European powers. It
has since completely lost its international character, and
become practically a Belgian dependency, under the rule
of King Leopold.

Extensive Annexations.—During the next few

years annexations proceeded apace in all directions.

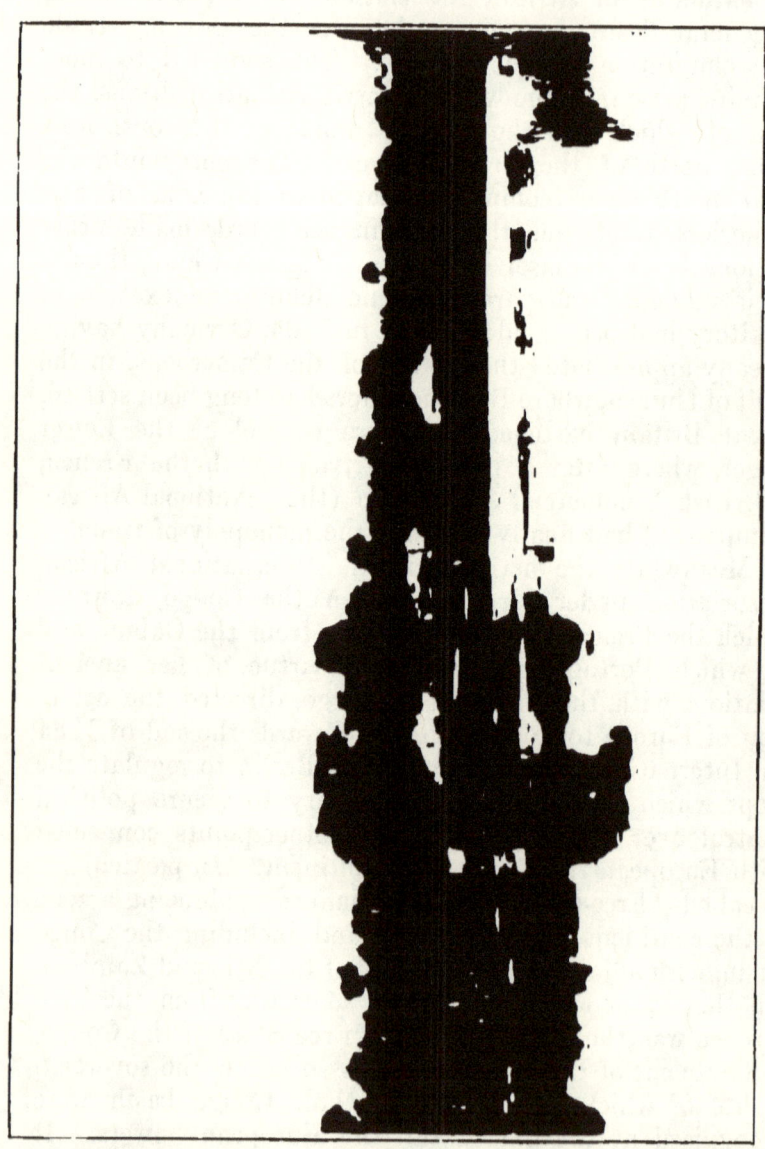

FIG. 10.—RABBA, ON THE NIGER, STATION OF THE ROYAL NIGER COMPANY.
(Photograph by Captain A. F. Mockler-Ferryman.)

Germany, by treaties with native chiefs, secured a footing inland from Zanzibar, in territory which had previously

been regarded as more or less under the rule of the Sultan.
Great Britain, too, as represented by Sir William Mackinnon and associated merchants, began to acquire influence in the country between Zanzibar and the Victoria Nyanza, and on the conclusion of an agreement with Germany delimitating the boundaries of the respective spheres, these merchants obtained a charter empowering them as the "Imperial British East Africa Company" to administer and develop the resources of the country. The territory remaining to the Sultan of Zanzibar was limited, besides the islands, to a strip of land ten miles wide along the coasts, which was subsequently leased to the Germans and English. On the Niger the National African Company extended its influence over Sokoto and Gando by treaties concluded by Mr. Joseph Thomson, and in 1886 obtained a charter granting increased powers as the "Royal Niger Company." An agreement with Germany secured to the Company both banks of the Benue as far as Yola. The coast-lands traversed by branches of the Niger delta and other "oil-rivers" (so called from the large production of palm oil) have been placed under the jurisdiction of the British Crown as the "Niger Coast Protectorate."

In South Africa, Bechuanaland had been definitely taken under British protection in 1884, and access to the Zambesi and Central Africa was obtained for British South Africa by a treaty concluded with Lo Bengula, king of Matabililand, in 1888, by which the whole of his dominions were placed under British protection. The administration of this territory was given by Royal Charter to the "South Africa Company." In Nyasaland, in spite of serious opposition from the Arab slave-traders, progress continued to be made.

About the same time **Italy** also joined in the scramble for African territory by establishing herself on the Red Sea coast, where the Bay of Assab, north of the straits of Babel-Mandeb, had been purchased in 1870, and proclaimed an Italian protectorate in 1880. When in 1885 Egypt, owing to the Mahdist rebellion, had been forced to evacuate her posts on the Red Sea, Italy stepped in and occupied the important port of Massaua, soon extending her occupation to the whole coast abreast of Abyssinia, with whose ruler, Menelek, a treaty was made in 1889. This ruler, however, has never acknowledged an Italian protectorate, and recent events have shown the difficulty of enforcing such on so formidable an antagonist.

The activity of **France** has been sustained from three different bases, viz. Algeria, the Senegal, and the Congo, where in 1885 the boundaries of her territories were greatly enlarged by an agreement with the Congo Free State, by which she gained access to the main river and the northern and western banks of its important tributary, the Mobangi. From these three bases French agents have converged on the interior, with the design of joining hands in the Central Sudan and forming a vast dominion embracing the greater part of West and West Central Africa. On the Senegal the military occupation was pushed on in the direction of the Upper Niger, on which gunboats were launched, and in course of time Timbuktu, the object of so many aspirations during several centuries, was reached and occupied.

International Agreements.—All these extensive annexations naturally led to conflicting claims by various nations, most of which were settled for the time by several important treaties concluded in 1890 and 1891, the result of which was to greatly extend the recognised boundaries of the various European spheres. Thus Great Britain obtained from Germany, in addition to the protectorate over Zanzibar, the recognition of her rights to Uganda and the Upper Nile region, to the region north-west of Lake Nyasa as far as Lakes Tanganyika and Mweru, and to the Lake Ngami region, while France agreed to the establishment of British influence on the Lower Niger and Central Sudan as far as a line drawn from Say, on the Niger, to Barruwa, on Lake Chad. From Portugal, after protracted negotiations, Great Britain obtained the recognition of her possessions on Lake Nyasa and in the central districts north and south of the Zambesi, including the Barotse country on the upper river. Germany obtained access to Lake Tanganyika for her East African possessions, and to the Zambesi for those in the

south-west. The French claims to the Upper and Middle
Niger, and the whole of the Central Sahara as far as
Lake Chad, were acknowledged by Great Britain (whose
West African colonies of Sierra Leone, the Gold Coast, etc.,
became thus debarred from advance into the interior), as
also were those of Italy to Abyssinia and the Somal and
Galla countries down to the Jub River on the east coast.
Finally, Spain obtained from France the acknowledgment of
her rights to a stretch of the Saharan coast north of Cape
Blanco, in the neighbourhood of which treaties had been
made with native chiefs, but little importance can be
attached to this strip of barren desert or to the small
portion of the coast north of the Gabun, surrounded by
French territory, of which Spain has also retained
possession.

These various agreements left many minor points un-
settled, some of which have been decided since, while some
are still under negotiation. The most important advance
has been that of France from her Congo territory in the
direction of the Upper Nile basin, and to Lake Chad
eastward of the German colony of the Cameroons, whereby
she has practically gained her object of joining her three
principal spheres in the centre of the continent. Many
of the boundaries thus fixed are extremely unsatisfactory,
in that they often follow arbitrary lines and take slight
account of geographical, ethnographical, or political facts.
They often cut across the courses of rivers and mountain
ranges, or separate portions of the same tribe or native
state. This may possibly be rectified in the future by a
modification of boundaries in accordance with natural
facts. By the recent partition of the continent but a small
area has been left unappropriated, and with the doubtful
exception of Abyssinia, the only native domains of any
importance are those of the Mahdists in the Eastern, and
of Wadai in the Central Sudan. A large part of the
Eastern Sahara, known as the Libyan desert, has not been
claimed by any Power, and as it may be regarded as
practically valueless, it is likely long to remain a "No
man's land."

CHAPTER V

NORTH AFRICA

1. *Physical Features*

LYING between the Mediterranean Sea on the north and the great desert of the Sahara on the south, the strip of land along the northern coast of Africa may be looked upon as a natural region, differing in many ways from the rest of the continent. The distinction is best marked in the western half, which has many points in common with South Europe. Farther east the country merges more gradually with the Sahara, while in Egypt, quite at the eastern end of the north coast, the valley of the Nile supplies a connection with more southern regions, and it will therefore be best to consider it along with the Nile countries as a whole. Apart from Egypt, North Africa falls naturally into two divisions—a mountainous western section, forming a block projecting beyond the general line of the coast, and a lower and more level eastern section. The west has a fairly plentiful rainfall, while in the east rain is much more scanty.

The Atlas Region. The mountain system of the Atlas fills up almost the whole of the western section. Its rocks are of later age than those of the main mass of Africa, and it has been formed, like many of the ranges of Europe, by a folding and wrinkling of the earth's crust— a structure which, on the whole, is not common in Africa. To the south the mountains sink to the moderately high plateau of the Sahara, the limit being marked in the extreme west by the valley of the Draa, and in the east

by a sunken area which stretches inland from the head of
the Gulf of Gabes. There is an important difference in
form between the western and eastern sections of the
system. In the west—in Marocco—it is a well-defined
mountain range, while in the east—in Algeria and Tunis—
it is rather a broad table-land with bordering ranges to
the north and south. The range of Marocco—sometimes
known as the Great Atlas—is by far the highest part of
the system, and forms a well-defined water-parting between
the streams flowing north and south. For a long distance
its average height is quite 10,000 feet, and the highest
peaks, several of which are over 13,000 feet high, are
covered for the greater part of the year, if not perpetually,
with snow. The highest known summit is the Jebel
Aiashi (14,000 feet), situated near the east end of the
highest section; but Mr. Joseph Thomson, who explored
a part of the range in 1888, considered that the peak of
Tamjurt, farther west, would probably prove higher. The
principal passes vary from 7000 to 12,000 feet, that of
Telremt, just east of Jebel Aiashi, which leads to the
oasis of Tafilet in the south, and is perhaps the most
frequented route across the mountains, being almost, if
not quite, 7000 feet. Parallel with the main chain in the
west of Marocco, there is a second of less altitude to the
south, which has been called the Anti-Atlas. It is broken
in places by streams which flow southwards from the
main range, but little is yet known of its special features.

In Algeria the northern border ranges, with the fertile
valleys which they enclose, are known collectively as the
"Tell." The principal ridges are the Jurjura uplands
(sometimes known as "Kabylia"), between the towns of
Algiers and Bougie, which rise to the height of 7570 feet
in Jebel Lalla; and the Sétif range, a little more to the
east, with Jebel Babor, about 6300 feet high. The table-
lands which fill up the central zone in Algeria are marked
by a series of salt swamps called "Shotts," flooded at one
period of the year. Similar temporary lakes also occur in
the Sahara at a lower level. The southern border range,
though less high and rugged than the Atlas of Marocco,
contains some groups such as the Jebel Amur and Jebel

Aures, with summits reaching to 6000 or 7000 feet. Jebel Sheliya (7600 feet), in the Aures group, is the highest peak in the part of the Atlas which lies in Algeria. In Tunis the mountains become lower and the plateau is broken up into ridges, which run down to the sea at Capes Bon and Blanc.

Streams of the Atlas.—There are no important rivers flowing down from the Atlas range; most of the streams have the character of mountain torrents during the rains, and almost, or completely, dry up in the dry season. They are therefore useless for purposes of navigation. On the southern slope especially the water-supply is precarious. The northern drainage of the Atlas of Marocco is gathered into four principal streams—the Tensift, Um-er-Rebia, Sebu, and Moluya. The Sebu, however, does not come from the main range, but from the lower hills to the north. The Um-er-Rebia runs for some distance parallel to the principal water-parting before turning north. The longitudinal direction of the streams is even more marked on the south side, the Wadi Sus flowing in its whole course along the hollow between the Atlas and Anti-Atlas; while the head-streams of the Draa likewise flow in the same longitudinal valley before breaking through the Anti-Atlas southwards. In Algeria the streams which rise in the outer slope of the mountains are quite insignificant, owing to the nearness of that slope to the sea. The principal river, the Shelif, rises on the inner plateau and flows on it for a long distance before breaking through the northern ranges. After passing through the inner range it turns westwards, and flows parallel to the sea for the second half of its course before finally leaving the mountains. Much of the drainage accumulates in the "Shotts" without ever reaching the sea. In Tunis the principal stream—the Mejerda—and another of less importance flow, like the Wadi Sus, at the opposite end of the Atlas, in courses generally parallel to the main axis throughout their whole length. The streams of the Atlas must formerly have been much more copious, even within historic times; and the Draa, in particular, may once have been an important river.

Climate and Vegetation.—The climate of the Atlas

region is in many ways like that of South Europe. Both the mean annual temperature and the range of temperature through the year are very much the same as in the Iberian Peninsula, owing partly to the effect of the sea breezes, and partly to the height of the mountains. South of these the temperature rises very suddenly. The amount of rain is also very similar, and, as in South Europe, most of it falls in winter, spring, or early summer. Partly on account of the similar climate, but still more because of a former land connection between the two regions, there is also a similarity in vegetation. The northern slopes of the Atlas are in part covered with forests of European trees, especially cedars and the cork and other oaks; elsewhere much of the surface is covered with low bush, made up of dwarf oaks or palms, olives, oleander, etc. On the drier southern slopes the vegetation is scanty. The plateaux have a steppe vegetation, with much *alfa* or esparto grass. The cultivated trees include the olive, fig, vine, and almond, and of cereals, wheat, barley, maize, and rice are grown. Other useful plants are the *Pistacia*, which produces *mastic* resin, an oak which supports the *kermes* insect, whence a scarlet dye is obtained, also the carob tree (*Ceratonia*), the pods of which (locust beans) are exported as fodder for cattle, besides hemp, flax, coriander, etc. The date palm is cultivated, but its natural home is in the desert regions to the south.

The **fauna** shows a mixture of South European forms, such as the fallow deer, moufflon or wild sheep, ichneumon, and porcupine, with other more southern types, such as the lion, leopard, hyaena, gazelle, oryx antelope, and others. Many of the birds of Europe migrate here during the winter, and this section of the fauna is, as might be expected, the most European in character. But besides specially European birds, such as the thrush, starling, and nightingale, many others, including birds of prey, the bee-eater, hoopoe, oriole, partridges, quails, the stork, pelican, and various kinds of ducks, occur. Reptiles are numerous, and locusts attack the country in swarms, doing immense damage to crops.

The **mineral** resources of the Atlas are considerable,

though perhaps hardly so great as was once supposed. The western part of the range especially is rich in iron and copper, and gold and silver, as well as other metals, are met with. Iron ore occurs also in the east of Algeria.

Tripolitana.—The rest of North Africa (excluding the Nile delta) is intermediate in natural characters between the Atlas region and the Sahara. The form of the land is characterised rather by extensive plateaux of moderate elevation than by mountain ranges. The strata—which include a greater extent of the newer rocks than is generally found in Africa—have not here been raised into a system of folds, as in the Atlas. The coasts, too, are lower, and show more recent deposits, which in places fringe the plateau escarpments by which the ancient shore line seems to be represented. Near the Gulf of Gabes (Little Syrtis) the plateau wall takes a semicircular sweep to the south, leaving a wide plain, called the Jefara, between it and the sea. The promontory of Barka, however, which bounds the "Syrtes" on the east, and is the end of an isolated plateau higher than the rest of the country, falls to the sea in a line of cliffs, and presents other resemblances to the Atlas. South of the "Syrtes" the interior consists in great part of a barren stony plateau or "Hammada," beyond which the ground sinks to the Fezzan basin, with its fertile oases. South of the Barka plateau, too, runs a long line of depression, occupied by the oases of Aujila, Jalo, etc., but little above sea-level. In the whole country there are no permanent streams of any importance.

The **climate** of this eastern region, owing to its nearness to the sea, is more moderate than that of the Sahara, but much less rain falls than in the Atlas, owing to the absence of high mountains. It occurs only during the winter months. The Barka plateau gets as much as 25 inches in the year, but elsewhere the amount is as little as 5 inches. The temperature varies immensely according as the wind is north or south, the mean for the year being about 70° or higher than in the Atlas. Poor steppes form much of the surface, and alfa grass abounds. In Barka, however, the trees of South Europe, especially the evergreen oak, arbutus, juniper, etc., flourish. A recent traveller has likened this region to the Malvern Hills or Shropshire, and speaks of glades of fine short grass and rich red soil. In the rest of the country the date palm grows naturally as far north as the coast, a fact which gives an indication of the general character of the flora. The wild animals are less numerous than in the Atlas region.

2. *Inhabitants of North Africa*

The Hamite Berbers, whose ancestors were the earliest inhabitants of North Africa of whom we have any knowledge, still form a large element in the population. Of the subsequent settlers and invaders—Phoenicians, Greeks, Romans, Vandals, and Arabs—the Arabs alone have left representatives in any great numbers, and these have so intermingled with the earlier inhabitants that no definite line of partition can now be drawn. In the northern coast-lands the Arab element is to be found chiefly in the east, though in the Sahara the reverse is the case. The Berbers occur in greatest numbers in the Atlas region, and are more inclined to settle down as cultivators than the Arabs, who still remain, as in their original home in Asia, a nomadic pastoral people. Both races have taken their part in the formation of states in the coast-lands, although in the Sahara, where the conditions are less favourable, little has been done by either in that direction. The states founded after the Moslem invasions have passed through many vicissitudes, but their mark is still to be seen in the political units of which North Africa is composed, which we have now to look at separately.

3. *Marocco*

Government, Limits, etc.—The word Marocco is the English form of Marakesh ("the adorned"), which is the Arabic name for the western capital of the country. The land itself was known to the early Arab geographers as Maghreb-el-Aksa ("the farthest west"). Our term, the "Moors," as applied to the inhabitants, is derived from their classical name *Mauri*, the meaning of which is not quite clear. This state is the only one of the old Moorish kingdoms of North Africa which has maintained its independence to the present day, having successfully resisted the Turkish encroachments in the sixteenth century, at which period it extended over a large area in the Western Sahara and parts of Algeria. The reigning dynasty, which dates from the beginning of the seventeenth century, was derived from the Hejaz in Arabia. Under Mulai Ismail (1672-1727) the highest degree of prosperity was reached, but the Empire has now fallen into decay, and European nations have of late years vied with each other in their efforts to obtain a footing. The present kingdom includes a section of the Atlas range, with the lower hills between it and the north coast, and also a part of the Sahara, but the Sultan's authority is there but small, and the region of Tuat, with its oases, is now falling under the influence of the French in Algeria. Only in the south-west is the southern boundary well defined by the course of the Draa.

The **population** of Marocco is generally estimated at about 6,000,000, the greater part of which occupies the northern slope of the country. Arabs and Berbers are much mixed up, but the Berbers as a rule inhabit the more hilly districts, while some of the lower plains are peopled by pure Arabs. The Berbers—mainly agriculturalists and traders—occupy fortified villages on the heights, each community being in great measure independent, but those belonging to the same tribe form a loose confederation under a *Kaid* nominated by the central authority.

Towns.—The towns of Marocco have quite an Eastern aspect, owing to the white quadrangular houses with massive walls, and countless minarets and domes of mosques. They may be divided into two groups: those on the coast, which serve as ports and centres of intercourse with the outside world, and those in the interior, which are the centres of government. Three of the inland cities serve in turn as the royal residence at different times of the year. Two of these, Fez and Mequinez, are in the upper basin of the Sebu, and the third, Marakesh or Marocco, in that of the Tensift. Fez lies at a considerable height between the northern spurs of the Atlas and the lower ranges to the north, and stretches for some distance along a narrow valley bounded by steep slopes. The population has been estimated at about

Fig. 11.—Great Gate, Mequinez.
(Photograph by Mr. Esme Howard.)

150,000, including a large number of Jews, who have a separate quarter here, as in all the large towns of the country. There is a Kasbah or citadel, and the whole is surrounded by a double ring of high walls. Gardens of fruit trees are to be seen both within and without. There are manufactures of silks, leather, and "fez" caps of red cloth. Mequinez (Meknes), which lies on a tributary of the Sebu, a little west of Fez, with which it is connected by a good road, was formerly much more important than it is now. Marocco has a population of about 50,000. It lies in a broad fertile plain resembling the bed of an ancient lake, from which the first terraces of the Atlas Mountains rise abruptly. Its mud walls, pierced by seven gates, enclose a vast area with many gardens and open spaces. It shows many signs of decay. Of the towns on the coast the most important are Tangier, on a sheltered bay backed by hills opposite the European coast: it was formerly occupied in turn by the Portuguese, Spaniards, and English, and is now the chief centre of foreign intercourse and outlet for the trade of Northern Marocco; Casablanca, with a good roadstead, which has lately become important as the port for Central Marocco; and Mogador, farther south, the principal outlet for the city of Marocco and the southern parts of the kingdom. All these are on the Atlantic coast. Within the straits several positions on the coast are occupied by Spain, the principal being Ceuta, just opposite to Gibraltar.

Resources and Trade.—The natural resources of Marocco consist chiefly in its animal and vegetable productions, for which both soil and climate are exceedingly favourable. Some of its plains are described as of matchless fertility, yet at present little advantage is taken of this. Both cereals and fruit trees (especially the olive) thrive well, and other crops include pulse, hemp, and flax, tobacco, etc. Sheep and goats and some cattle are reared, and the skins and wool support the industries of leather-work and weaving (including that of rugs and carpets), for which the country has long been famous, besides supplying exports to foreign countries. Besides the trade by sea with Europe—still very small—a certain amount is carried on by means of camels with the regions south of the Atlas, as far even as the Sudan. Dates are also brought from the oases south of the mountains, especially Tafilet, and some quantities are exported to Europe. As most of the goods are carried on the backs of camels, good roads hardly exist, and though so near Europe, railways have not yet been introduced into the country.

4. *Algeria and Tunis*

These two countries, situated to the east of Marocco, on the northern coast of Africa, were long nominally vassal states of the Turkish Empire, though in great measure independent under powerful rulers of their own, known as Deys of Algiers and Beys of Tunis. Algiers was the headquarters of the "Barbary Corsairs," long the terror of the Mediterranean, whose piratical proceedings led to the bombardment of the town by the English and Dutch in 1816. Algeria was conquered by the French in 1830 and following years, but Tunis remained under the old *régime* until 1881, when the Bey acknowledged the French supremacy.

Algeria occupies, besides the mountains and plateaux of the Atlas range, a large area of the Sahara to the south, but the boundaries there are ill-defined. As in Marocco, the bulk of the population consists of Arabs and Berbers,— now all bigoted Mohammedans,—but these have been supplemented of late years by a large influx of Europeans, who form about an eighth of the total population of 4,000,000. Of these the largest number are French, and after them Spaniards and Italians are the most numerous. Algeria sends representatives to the French Parliament, but the administration is in the hands of a civil governor (assisted by a council), who has taken the place of the military authority, under which the country was long placed. It is divided into three departments, corresponding almost exactly with the three subdivisions in the time of the Deys. Each occupies a section of the country from north to south, including a portion of the Sahara, which, however, still remains under military rule. The French are gradually extending their authority southwards, and now lay claim to the oases of Tuat, which used to belong to Marocco. All attempts to advance into the region of the hostile Tuareg of the Sahara have, however, so far failed.

Towns.—The greater part of the population occupies the northern zone or "Tell" region, the largest towns being found on the sea-coast, owing to the nearness to it of the outer terraces of the Atlas system. A few are placed on the plateaux which succeed the Tell, but the southern parts of the country are very thinly peopled. A large part of the population is scattered in small agricultural communities, so that none of the towns are very large. The coast towns lie, as a rule, in bays sheltered by headlands which run out from the interior highlands. Algiers (82,000), on the west side of the spacious and sheltered bay of the same name, is built on the lower terraces of the coast range, the green slopes of which rise in the background above the glaring white houses of the town. It is strongly fortified, and has both a commercial and a military harbour protected by breakwaters. The European quarter, with its fine *Boulevard de la République*, occupies the shore and the lowest terrace, and behind it is the Arab town, with its narrow streets surmounted by the citadel or Kasbah. Oran, the chief town of the west, little smaller than Algiers, is placed in a somewhat similar position, but it is not so well sheltered by nature. A breakwater now forms an artificial harbour. It is also strongly fortified. Another western port is Mostaganem, near the mouth of the Shelif River. The chief ports of the eastern department are: Bona, near the eastern frontier, the most European town in Algeria, with a fine promenade dividing it in two halves; Philippeville, and Bougie. The chief town of the department, Constantine (41,000), lies away from the sea at the northern edge of the plateau of the Shotts, Philippeville serving as its port. It occupies the site of Cirta, the ancient capital of Numidia, on an eminence overlooking the gorge of the Wadi Rummel, and was rebuilt under its present name by the Emperor Constantine. It was taken by the French in 1837. The other interior towns are mostly small, although Roman and other ruins are scattered over the whole country, showing how much larger the population once was. The chief are: Tlemcen, Sidi-bel-Abbes, Blidah, and Guelma in the "Tell," Géryville and Batna on the interior plateau, and Biskra on the southern slope, facing the Sahara, in the midst of luxuriant palm groves. Tlemcen, near the frontier of Marocco, was in former times an

Fig. 12.—Street in Constantine.
(From print by Gervais, Courtellemont, and Co.)

important centre of Berber civilisation and trade, and then contained a population many times larger than at present. In the Algerian Sahara the principal centres of population are the oases of Wargla and Tuggurt (on the course of the old river-bed of the Wadi Mia) and El Golea, farther to the south-west. Cultivation (and with it population) has of late years been much extended in this part of Algeria by the boring of Artesian wells, which now irrigate many districts formerly arid.

Industries, Trade, and Resources.—By far the largest part of the population of Algeria is engaged in agriculture. This forms the occupation of nearly half the foreign population, as well as a large proportion of the natives. Besides cereals (including wheat), the vine is now largely grown, and the area devoted to it has enormously increased within the last twenty-five years. Forests of evergreen oaks and conifers still cover parts of the "Tell," but much has been destroyed by the Arab occupiers. There are few manufactures, but leather-working is carried on at Constantine. The country is rich in minerals, copper and iron ores being exported. Other exports are alfa grass (used for making paper), which grows in great quantities on the plateaux, and garden produce, which is sent to France, England, and Germany from the neighbourhood of Algiers. Since the French occupation trade has much increased. It is carried on chiefly with the mother country.

Communications.—The "Tell" is now fairly well provided with railways, which afford communication between the ports and interior towns, and between the towns of the different departments in a direction parallel to the coast. Owing to the nearness of the first terraces of the Atlas to the sea, the railways in this direction mostly make use of the valleys of streams running behind the outer ranges. Thus the capital is connected with Mostaganem and Oran by a line following the east to west course of the Shelif, and with Bougie and Constantine by one running behind the Jurjura highlands. Constantine is connected by railways with its port of Philippeville, with Bona, and by the valley of the Mejerda with Tunis. The plateaux are, on the whole, poorly supplied. The roads are bad, and only in the extreme east and west do

railways traverse it from north to south. In the east a
line leads to Biskra by El Kantara, the gate or pass by
which one of the most frequented routes crosses the southern
border ranges of the Atlas, and in the west another has
been carried to Ain-Sefra, near the frontier of Marocco,
but so far there is no line from Algiers southwards, in which
direction another trade road crosses the mountains to
Laghuat. All these southern routes, including some from
Marocco and Tripoli, converge on Tuat and Tidikelt,
which thus are of great importance as commanding the
road to Timbuktu. Railways will probably be carried in
time to the southern oases, even though the projected line
across the Sahara should never be carried out. In spite
of the antiquity of the trade with the Sudan across the
desert, its proportions are not large, and the access opened
to the Sudan by way of the Niger has already proved
prejudicial to it.

In spite of the natural advantages of Algeria, its
economic condition is far from favourable, owing mainly
to the expenses of the military occupation, necessary to
keep in check the disaffected Mohammedan population.

Tunis occupies the eastern end of the Atlas region,
where the last ridges of the range plunge beneath the
sea, causing the sudden southerly bend of the coast.
Although it has been a French Protectorate since 1881, a
Bey still nominally rules, but the actual power is in the
hands of the French Minister, who is supported by a
strong military garrison. The native population is better
disposed to foreigners than that of Algeria, although at
the time of the French occupation the tribe of the Krumir
in the west made a stubborn resistance. The total population is about 1½ millions, the foreign element being much
smaller than in Algeria. Italians and Jews form the most
numerous sections of it. The chief town, also named
Tunis, stands at the head of the shallow bay of the same
name, and is, next to Alexandria, the most important
town of the north coast of Africa, although devoid of a
good harbour. Goletta, which serves as its port, has
merely an open roadstead. The position of Tunis,

commanding the most central and narrowest part of the Mediterranean, long ago brought this spot into prominence, for here Carthage once stood, and its scanty ruins are close to the modern city. This has remained until lately a true Oriental city, with narrow dirty streets and little to attract the senses. Bizerta, farther north, stands in a position more favourable for shipping, and is likely to become a strong naval arsenal. The harbour has lately been improved by the digging of a canal leading to an inner basin. Other ports are Susa and Sfax, on the gulfs of Hammamet and Gabes respectively, which indent the east coast. The most celebrated inland town is Kairwan, a sacred city of the Moslem world, from which unbelievers have until lately been rigorously excluded. It has a very fine mosque.

The **products** of Tunis are similar to those of Algeria, but cultivation has not yet reached the same extent, owing to the recent date of the French occupation. The soil is said to be more fertile than that of Algeria, and with an increasing trade the financial prospects are better. The capital is connected with the Algerian system of railways by the valley of the Mejerda. A project which was formed for letting in the waters of the sea over the depressed region of Shotts, at the head of the Gulf of Gabes, has now been abandoned.

5. *Tripolitana*

Alone of the Turkish dominions in North Africa, Tripoli, which occupies the space between Tunis and Egypt, has remained an integral part of the Turkish Empire, of which it forms a province, both the civil and military affairs being administered by a *vali*. The ancient Berber population has here become almost completely fused with the Arab invaders, and there is but a small sprinkling of Turks, Jews, Italians, etc., in addition. The total is estimated at about a million. The province of Tripoli includes the oases of Fezzan to the south. The only important town on the coast is the capital, Tripoli (10,000), in which almost all the trade of the country centres. It lies at the edge of the plain of the Jefara. Seen from a distance it presents an imposing appearance, with its castle and minarets, but the impression is less favourable on a nearer view. As the terminus of one of the oldest and most important trade routes across the Sahara, it has played an important part in the opening up of Africa, many of the great exploring expeditions having made it a starting-point. The principal towns of the interior are Murzuk in Fezzan, on the direct southern route across the Sahara, and Ghadames, near the frontier of Algeria, on the route to Tuat and Timbuktu. The port of Barka is Benghazi, whence a caravan route leads southwards by the Aujila oasis and that of Kufra to Wadai. Besides these southern routes there is one leading along the length of the country from Tripoli to Egypt by the longitudinal depression of Aujila and Jalo.

The only important product of Tripoli is the *alfa* grass, the export of which has largely increased of late. The exports include some products of the southern countries brought by caravan, such as ostrich feathers, ivory, and gold, but the value has much diminished, except in the case of ivory, the trade in which has been stimulated by the Mahdist revolt, which has blocked the export route from the Sudan by way of Egypt.

CHAPTER VI

THE SAHARA

1. *Physical Features*

General Character.—The whole breadth of North Africa, from the Atlantic on the west to the Nile on the east, is occupied by the region known as the Sahara, from the Arabic "Sahrā," an uninhabitable waste. In spite of great variety of forms of surface, almost the whole of this vast area has this common characteristic, that it has a very scanty supply of water, and therefore is generally either completely wanting in vegetation, or only slightly covered with a few plants which have adapted themselves to the arid climate. It forms the western end of a great zone of deserts, which stretches almost across the entire width of the Old World, from Mongolia in North-East Asia through Turkestan, Persia, and Arabia to the shores of the Atlantic. It is, however, divided off from the rest by the Red Sea and Nile valley, and forms in itself the largest continuous desert on the earth's surface, its area being about $3\frac{1}{2}$ million square miles, or nearly equal to that of all Europe.

Soil and Relief. The idea formerly prevailed that the whole surface was one vast sea of sand of almost unbroken monotony, and it was generally held that it had been quite recently covered by the sea, which had retreated, and left its floor uncovered. But now that explorers have traversed it in many directions it has been found to have as great a variety of soil and relief as any other part of the Earth's surface. Although, taken as a whole, it may be described as a plateau of moderate elevation, this does not mean that it is one unbroken level, for it is traversed by some fairly high ranges of mountains. In particular, a line of high ground branches off from the high southern plateau of Africa and runs north-west across the whole width of the Sahara, almost to the shores of the Mediterranean, while a second runs from south to north between the basins of the Niger and Lake Chad, and joins the first in the extensive high plateau of Ahaggar, south of Algeria. Between these two lines of high ground and the high southern plateau the basin of Lake Chad is completely enclosed on all sides. The highest parts of these ridges form groups of serrated mountains rising to a height of 6000 and 7000 feet above the sea, the principal being those of Tibesti in the eastern and Air in the western line. A common feature, especially in the north central parts, is the occurrence of isolated plateaux with rock-strewn surfaces, and bounded by nearly perpendicular cliffs. The eastern and western sections of the Sahara are generally lower, and there, especially in the eastern section, which forms the Libyan desert, the old idea as to the nature of the country is more borne out. In the west there is a depression named El Juf, only slightly raised above the sea.

There is an equal variety also in the structure of the country, the mountain ridges, which, as it were, form the skeleton of the whole, being composed of ancient crystalline rocks (sometimes with a core of volcanic rocks thrust up from below) flanked by ancient sedimentary strata, while the lower grounds are made of newer formations and largely covered with loose sand.

Causes of the Desert.—The desert nature of the Sahara is not due in the first instance to any peculiarity of the soil, but merely to the absence of sufficient moisture to support vegetation. The main reasons for this want of rain have already been stated in the first chapter. In winter the winds blow chiefly outwards from the Sahara towards the surrounding regions, while those which in summer blow towards the interior, owing to the greatly heated surface of this vast mass of land, become themselves more and more heated, and tend to evaporate more moisture, instead of parting with it as rain. The Atlas range also does its part in drying the air by condensing the moisture of the winds which blow over it from the north-west. If there were more high mountains in the centre of the country no doubt there would be more rain, for in the mountainous districts of Air and Tibesti local rains of some violence occur during one part of the year, and cause sudden torrents to rush down to the lower grounds. Sudden storms also occur in summer away from the mountains, especially towards the southern edge of the desert, where the southerly winds blow from a hot and moist region. It seems likely that in former times the climate was wetter than at present. This is shown by the ruins of buildings erected by former inhabitants and by the courses of streams which are now only occasionally filled with water. The high plateaux south of Algeria must once have been the centre from which the drainage of the whole region diverged. The old rivers are now represented by the Wadi of the Igharghar, which has a northward direction to the lowlands of the Shotts in the south of Algeria, and by a system of dry water-courses directed southwards towards the Niger basin, or westwards towards the region of sands which occupies the Western Sahara.

Surface Features.—The aspect of the Sahara is being constantly changed by the wearing away of the surface, although this is of a different kind from that which acts in moister regions. Owing to the sparseness of vegetation the sudden torrents which occasionally rush down the dry beds of Wadis have an unusual power of tearing away the loose soil. Another active agent in wearing down the surface is the enormous variation in temperature between the day and night, by which the outer layers of the rocks are expanded and contracted so rapidly that they split in all directions, and some of the plateaux are simply strewn with angular fragments of rock. The plateaux themselves are being constantly worn away by the splitting off of large fragments from their steep scarps. The ruins, so to speak, of the former surface often form chaotic accumulations of rocks without any definite arrangement, known by the name "gur." The sand which results from this breaking up of the rocks is carried hither and thither by the wind, and by its scouring action does its part in the general work of destruction. Violent sand-storms occur, in which the air is so filled

Fig. 13.—Sand-dunes in the Sahara.

with sand, that a sudden darkness ensues, and travellers run the risk of being suffocated.

Over vast areas, called Erg or Igidi, the sand is piled up into dunes or hills almost entirely devoid of vegetation, which, like waves of the sea (though much more slowly), are constantly shifting their position, often encroaching on fertile oases, and burying everything under their barren surface. Their height sometimes reaches 600 feet and more. Two great areas of dunes occur in the east and west of the Sahara respectively. The Libyan desert is one of them, and the other occupies a vast extent south of the Atlas, on the southern borders of Algeria and Marocco. By these the ancient relief of the country, which may have been formed when the climate differed from the present, is gradually changed, the lower parts of old river-beds being often blocked up and the slope completely altered. Enclosed basins may be thus formed in parts where streams once found their way to the sea. These shifting hills of sand, however, serve one good purpose in that they help to screen the scanty moisture that falls in parts of the desert, from the intense evaporation. Sinking below the surface, the water finds its way for long distances underground, so that by the sinking of wells it can be obtained in parts where the surface is absolutely arid and bare. In this way oases are formed, whether along ancient watercourses, beneath which there is often a buried stream of water, or in other parts, such as in the Libyan desert, where the water is collected in hollows of the impervious underground strata.

The rocky plateaux or "Hammada," on the other hand, on account of their elevation and porous composition (being largely formed of limestone rocks), are not so well supplied, so that these are in reality the most absolutely desert parts of the whole region. The deepest hollows, in which the old river-beds often end, are occupied by Shotts and Sebkas, the representatives of ancient lakes now dried up by evaporation. According to the amount of water, these are alternately salt swamps or level expanses of dry salt glistening like lakes in the sun. The amount of salt is sometimes so great that it forms a crust like ice on the surface of the water.

Flora and Fauna.—The plant-life of the Sahara consists of two sections, the true desert plants and those of the oases. The desert plants have in time become specially adapted to resist the extreme dryness, and form a well-marked group, differing from those of neighbouring regions. They are mostly of a stunted or tufted appearance, growing at wide intervals over the surface. The leaves are either exceedingly minute or so leathery that there is little evaporation from their surface, and many of the plants bristle with thorns, which seem to take the place of leaves. The principal are the gum-acacia (*Talha*), the tamarisk (*Ethel*), one or two of the pea family of plants, such as the *Agol* and *Retem* (the latter allied to the broom), and a species of *Ziziphus*, a plant with a sweet, plum-like fruit, found throughout North Africa. The date palm is the principal tree of the oases, and is the most important product of the Sahara, for it supplies the staple food of all the people, and without it life would hardly be possible. It occurs in groves or extensive thickets. Though introduced only after the commencement of the Christian era, the camel is now the domestic animal *par excellence* of the Sahara, but in the more favoured parts the horse, goat, and sheep are found. The wild animals are few. They include the lion (met with in a few spots), the leopard, jackal, hyaena, fox, gazelle, wild boar, etc.; the giraffe is found in the south-east parts. Of birds the ostrich is the most important.

2. *Inhabitants of the Sahara*

The races of the Sahara show signs of great mixture. It has been supposed by some that the Negro race was once spread over this region, but was driven out by the Hamite intruders, and the dark hue of certain of the tribes is thought to be indicative of a mixture with the Negro. It is, however, chiefly in the oases, to which Negro slaves have for centuries been brought from the south, that the darkest individuals are found.

Three main divisions of the Saharan peoples may be distinguished: the Arab tribes in the west, the Tuareg or Imoshagh, probably Berbers, in the centre and west centre, and the Tibu or Teda in the east. The relationship of these last is very doubtful, some considering that they are connected with the Negroes, and others that they are a Hamite people distinct from the Berbers. The dividing line between the Tibu and Tuareg falls about the meridian of Tripoli, but a great part of the Sahara east of this is completely uninhabited, and the Tibu are limited to the Tibesti highlands and neighbouring districts. The Tuareg and Arabs are not separated by any definite line, as the western Tuareg have in many cases adopted the language, religion, and customs of their neighbours. Physically, the desert tribes are marked by great leanness, but are otherwise well-made and capable of enduring fatigue in an unusual degree. The general dryness of the air makes it exceedingly healthy, and long life is common. Only in the moister oases and on the borders of the Sudan do the inhabitants suffer from malarial fevers, and in these places the people are of a much feebler type.

Modes of Life.—Two broad classes may be distinguished among the Saharan peoples: the nomads and the fixed population. The nomads may be considered as the typical desert inhabitants, while the fixed population is found only in the cultivable oases, and is the most mixed with foreign blood. Scanty pasturage for flocks being the only product of the greater part of the surface, most of the tribes (Arab and Berber alike) are of necessity pastoral, and obliged to wander from place to place in their search for fresh feeding-grounds. While the Northern tribes possess horses, and those on the southern edge of the desert rear goats, or even cattle, those in the central zone depend above all on their camels, which supply them with the only possible means of transport, as well as with milk, cheese, and occasionally flesh.

The roving life of the nomads is naturally unfavourable to complex social or political organisation, and tends to split up the tribes into small family groups under patriarchal rule, each of which is unwilling to recognise any outside authority. Yet there are loose confederations both among the Tuareg and Arabs. Among the Tuareg, those of the Ahaggar and Azjer of the plateaux south of Algeria, Kel-Ui of Asben, and Awelimmiden, north of the Middle Niger, are the most important; and among the Arabs, the Ulad-Delem and the Ulad Bu-Sba in the Western Sahara, and the Ulad Sidi esh Sheikh along its northern borders. These nomad tribes can with difficulty find sufficient means of subsistence, and are therefore much given to plunder, and are greatly feared by the inhabitants of the oases.

These last live by cultivating the soil, which they are able to do by the help of various methods of irrigation, either from wells or by channels derived from running streams. Except in the south of Algeria, where artesian wells have been introduced, the cultivation of the oases is declining, and the population therefore decreasing, as the supply of food does not meet the requirements, while blackmail has often to be paid to the nomads of the district. The most important culture in the oases is that of the date palm, but cereals are also grown, as well as vegetables, tobacco, cotton, and some of the fruit trees of the Mediterranean.

In addition to the oases which have been mentioned as belonging to the countries of North Africa, the principal centres of population are the mountainous districts of Tibesti, Borku, and Air or Asben, with the oases of Bilma and Kauar, west of Tibesti, in the central region; the oases of Kufra, in the Libyan desert; and Adrar, in the west.

Trade and Trade Routes.—The introduction of the camel into North Africa in the early days of the Christian era has provided a means of communication across the desert between the peoples of the Mediterranean sea-board and those of the fertile regions of Tropical Africa. The products of these two regions being quite different, the desire for an interchange of commodities naturally sprang up. The main products of the south are gold, slaves, ostrich feathers, ivory, gums, and wax, and at the present day those carried in exchange from the north are principally English and German cottons, sugar, tea, and miscellaneous small articles. Besides this transit trade, the two staple products of the desert, dates and salt, form articles of trade between the desert and the countries north and south. Dates are exchanged, principally by the inhabitants of the northern oases, for the products of the Atlas, consisting of live sheep, dried meat, wool, butter, cheese, and wheat, which are annually carried southwards by the inhabitants of the Atlas countries. Towards the south, salt (obtained chiefly from Taudeni, north of Timbuktu) is exchanged in like manner for the products of the Sudan, such as rice and millet, cotton and slaves. This traffic is carried on by means of regularly organised caravans, often numbering hundreds of camels, the supply of which becomes a regular business with the desert tribes. The need of a water-supply leads to the choice of certain fixed routes passing through the principal oases or scattered wells in the midst of the sea of sand, and determined also by the character of the different tribes. According as the kingdoms south of the Sahara have risen into importance or declined, different routes have come into prominence at different epochs. At the present day five are in use, passing respectively through the five important centres (from west to east), Tenduf, Tuat, Rhat, Murzuk, and Kufra, all lying in nearly the same latitude near the northern border of the desert. North of these points the routes branch or interlace to some extent. The most westerly route leads from Mogador in Marocco to Timbuktu, and the most easterly from Benghazi in Tripolitana to Wadai. The Tuat route, connected northwards with Marocco, Algeria, and Tripoli, also leads southwards to Timbuktu, while the two remaining lead from Tripoli—the one (*via* Rhat and Air) to the Hausa towns of Sokoto, Katsena, and Kano, the other (*via* the Kauar and Bilma oases) to Kuka, the capital of Bornu. Thus all the three central routes lead in one way or another to Tripoli, which is one of the most important termini of the commerce across the Sahara. The most central route of all, which leads to Kano, the great trade centre of the Central Sudan, is perhaps the most frequented. Those to Kuka and Timbuktu have lost much of their former importance, but the route between Wadai and Benghazi has come into prominence only during the nineteenth century. No European traveller has ever followed it southwards beyond the Kufra oases, which Rohlfs visited in 1879. Between these routes lie the most inhospitable stretches of sand, such as the sand dune region south of Marocco, which is almost deserted even by Arab and Tuareg nomads.

The trade of the Sahara is of small amount, judged by modern standards, and has been declining since the opening of the Sudan to trade by the sea routes, although it is said that that of Timbuktu has revived since the French occupation.

Political Relations.—While the Libyan desert to the east, on account of its hopeless barrenness, has not been claimed by any outside power, the central and western regions of the Sahara have in great measure fallen within the "spheres of influence" of France and Spain. By international agreements France has been recognised as the dominant power over the whole Central Sahara from Algeria to the borders of the Sudan, and also holds the Atlantic coast north of the Senegal as far as Cape Blanco, Spain claiming the coast between Capes Blanco and Bojador. No boundaries have been fixed between these spheres in the interior. So far the influence of these countries in the central parts exists in mere name, the nomad tribes still retaining their independence, owing to the protection afforded by the desert against outside aggression. The French project of a railway across the Sahara to unite Algeria with the Sudan can never be a commercial success, though its strategical importance would be considerable.

CHAPTER VII

THE SUDAN

1. *Physical Features*

General Characters.—As the rainfall increases along the southern margin of the North African deserts, the nature of the country gradually improves, until a more fertile zone is reached, which stretches across the continent from the Senegal to the Nile, and reaches southwards to the Gulf of Guinea and the beginning of the high table-land of Central and South Africa. In this zone the inhabitants also change, and the Semitic and Hamitic races belonging to the white division of mankind give place to the black or Negro race. On account of the dark colour of the people, this region became known to the dwellers in North Africa as "Bilad-es-Sudan," or the Country of the Blacks, which term, shortened to "the Sudan," is now generally used to describe the zone bordering the desert on the south. As the eastern portion of this zone belongs naturally to the region of the Nile, we shall deal in this chapter only with the central and western parts. This vast region falls naturally into two main subdivisions, very different in physical character, but so connected politically that it is necessary to consider them together—an interior plateau of moderate elevation, consisting largely of open grassy savannahs, and a low lying strip of country stretching along the coast of the Gulf of Guinea, in great part forest-clad. With the plateau region we may include the lower grounds watered by the Senegal and Gambia, which have more in common with it than with the Guinea coast-lands.

The Zone of Savannahs.—The general character of the interior plateau of the Sudan is that of a gently undulating plain, from the surface of which isolated ridges and summits rise. These for the most part consist of granite, or other ancient crystalline rocks, which seem to form the framework of the whole region, though covered in parts by later, but still ancient, sedimentary strata, or recent alluvial deposits. According to the river-systems this region may be divided into three sections : (1) the country watered by the Senegal and Gambia, (2) the Niger basin, (3) the basin of Lake Chad.

The **Senegal** and **Gambia** both rise in the mountainous country of Futa Jallon, but the course of the Gambia to the sea is the more direct, and therefore the shorter of the two (about 600 miles). Both streams are broken by rapids before reaching the lowlands, but owing to its greater length (950 miles), the Senegal has longer stretches of navigable water. Its principal head-stream, the Bafing, flows east and north before taking a decided westerly direction near its junction with a second branch, the Bakhoi, which comes from the south-east. Soon after the junction the stream is broken by the Guina and Felu falls, which are the limit of navigation on the lower river. Farther on, the Faleme (also from the Futa Jallon highlands) joins it on the south. In its lower course the Senegal sends off several sluggish side-channels, which unite with the main stream during the rains to form a continuous sheet of water. Its mouth is barred by a long line of shifting sands, which bend it southwards, forming a kind of lagoon parallel to the coast. The Gambia flows north-west for about 150 miles, but in spite of many windings, its general direction is then west to the sea, before reaching which it forms a wide estuary. At about the middle of its course it is broken by the Barrakunda rapids, which prevent vessels from reaching the upper stream from the sea. The lowlands through which the Senegal and Gambia flow are not clothed with dense forests, and some of the country has almost the character of the desert, which begins on the north bank of the Senegal.

The **Niger**[1] basin is, for the most part, an undulating plain, broken by few elevations of any importance. Even round its outer edge the higher grounds which separate it from the coast stream on the one hand, and from the Lake Chad basin on the other, hardly anywhere take the form of regular mountain ranges. The most important groups of mountains occur at the south-west and south-east corners of the basin, giving rise to the main stream and its principal tributary respectively. The highlands in the south-west, together with those of Futa Jallon, whence spring the Senegal and Gambia, rise to a height of 4000 to 5000 feet. Those to the south-east are still higher, but belong properly rather to the high plateau of Central Africa than to the Sudan. The southern limit of the Niger basin, formerly supposed to consist of a well-defined range which went by the name of the Kong Mountains, has been found to consist of a strip of plateau, with a steep escarpment towards the Gulf of Guinea, and with some isolated peaks rising above the general level. The water-parting between the streams flowing north and south is a curved line following the highest crest of the plateau. In the centre of the basin the only mountains deserving of mention are those of Hombori, a group of fantastically shaped peaks resembling columns and castle walls. The main stream of the Niger rises in the forest-clad hills in the south-west corner of the basin, only 140 miles from the coast, and flows north-east, or exactly away from the sea, for over 1000 miles, during the latter part of which it traverses a region of swampy lakes and backwaters, of which Lake Debu is the chief. Then bending round east and south-east, it describes a vast semicircular curve, and after receiving its great eastern tributary, the Benue, enters the sea near the head of the Gulf of Guinea by a large swampy delta, in which it becomes split up into a number of branches. Except the Benue it receives few tributaries. Most flow from the southern plateau, but some also from the east, of

---

[1] The name Niger is not known to the natives, but has been borrowed from the ancient writers. In the country itself the names used most frequently are Joliba for the upper, and Kwara for the lower part of the river.

which the chief is that flowing past the city of Sokoto. To the north the desert approaches close to the main stream at its northern bend, and the limits of the basin are ill defined. The average volume of water brought down by the Niger is second only to that of the Congo among African rivers, but in its length and the size of its basin it falls short of the Nile. Both the Upper and Lower Niger are navigable for hundreds of miles, but the river is broken by rapids in the centre of its course.

The basin of **Lake Chad** is surrounded on all sides by higher ground, so that none of its water ever reaches the sea. Lake Chad, into which the various streams converge, is not quite the lowest part of the basin, for a greater depression that of Bodele—exists to the north-east, into which some of the surplus water finds its way in unusually wet seasons by the channel of the Bahr-el-Ghazal, which, however, is, as a rule, dry, falling as it does within the limits of the arid region of the Sahara. Lake Chad is a shallow, swampy lake, its shores fringed with reeds, studded with islands, and varying immensely in size according to the season of the year, shrinking in the dry season to less than half its full extent. Its main feeder is the Shari, which, with its various branches, drains the country on the south and east, its main tributaries rising on the northern slope of the Central African plateau. The only other feeder of Lake Chad is the Waube, which drains the country on the west, and brings down a considerable amount of water in the rains, though in the dry season it shrinks to a small size.

Climate, Flora, and Fauna.—In this zone the climate and the character of the country vary according to the distance from the Sahara, the rainfall increasing and the soil becoming more fertile and better cultivated the farther it is from the sandy desert. On the northern borders sand-dunes occur. Farther south bare or grassy stretches, traversed by granite ridges and sprinkled with acacias, tamarinds, dum palms, etc., alternate with cultivated areas or patches of wood in moister spots. More extensive forests occur near the southern borders. The corn most cultivated is durra or Sorghum (a grain somewhat resembling millet), which, though a native of the East Indies, is *par excellence* the corn of Africa. Cotton is also much grown, together with beans, rice, indigo, and various vegetables. The wild animals include most of the forms found generally in the savannah regions of Africa (elephant, rhinoceros, hippopotamus, crocodile, lion, leopard, antelope, etc.). The giraffe is not found in the western parts. On the southern borders the fauna becomes more like that of the coast-lands, including, *e.g.* the chimpanzee.

The Coast-lands.—The coast-lands, to which the name Upper Guinea is often given, are distinguished from the interior regions principally by the lower elevation and greater humidity, which in turn act on the vegetable and animal productions. They show many resemblances to the more southerly parts of West Africa, including the Congo basin. They occupy the interval of varying width between the escarpment of the interior plateau and the sea, in which the ground, as a rule, rises very gradually by one or more steps of slight elevation. A distinction may be drawn between the coasts facing westwards or south-westwards and those facing southwards, Cape Palmas forming the point of separation between the two. East of this cape the shore is more sheltered from the Atlantic gales, which blow with full violence on the coast between Cape Palmas and the Senegal. Again, east of this point the coast is marked by the presence of a system of lagoons, separated from the sea by a narrow strip of sand or mud, which are wanting to the west. The combined influence of the eastward-flowing Guinea current and the ceaseless surf of the Atlantic piles up the sand, damming back the coast streams, which become choked with growths of mangroves, papyrus, and reeds, and lose the power of clearing out a channel to the sea. Thus the old coast-line becomes separated from the sea by a belt of low swampy country, in which not a single elevation is to be seen from a ship's deck. This formation occurs chiefly in the incurves of the coast, especially in the Bight of Benin, west of the Niger delta, the more projecting parts being higher and firmer ground. As we advance into the interior the swampy zone is succeeded by a zone of forest, which gives place to more open country, broken by granite and other rocks, when the plateau escarpment is reached.

Different parts of the coast have been known—since the first opening up of trade by European nations—by names derived from the various products for which they were noted. These are the Grain or Pepper Coast, west of Cape

Fig. 14.—Lagoon, with Mangroves.
(Photograph by Rev. J. T. F. Halligey.)

Palmas; the Ivory Coast, occupying the incurve between Cape Palmas and Cape Three Points; the Gold Coast, between Cape Three Points and the Volta, and the Slave Coast, east of that river. These divisions correspond in some degree to natural subdivisions of the coast. Thus the Ivory and Slave Coasts are fringed with lagoons, while the Gold Coast is mostly composed of steep wooded cliffs.

Rivers.—The rivers of this region have mostly short courses rising in the plateau escarpment and flowing direct to the sea. Those towards the north-west have mostly broad estuaries quite out of proportion to their length or body of water. Such are the Casamanza, Geba, Rio

Grande, and Scarcies. The rivers of the Slave Coast and Bight of Benin have also short courses, but the Volta, on the Gold Coast, and the Akba (Komoe) and Bandama, on the Ivory Coast, have been found by recent explorations to have their sources far north on the uplands of the interior. The Volta is the principal river of the whole coast (excepting, of course, the Niger), its basin extending beyond 12° north latitude

Climate, Flora, and Fauna.—In these respects Upper

Fig. 15. River Ancobra, Gold Coast.
(Photograph by Rev. J. T. F. Halligey.)

Guinea forms part of the same natural region as the Congo basin and adjacent coasts, much of the surface being clothed with forest, the growth of which is favoured by the moist and equable climate, deadly as it is to white men. The characteristic tree is the oil palm, whose fruit supplies palm oil, an important article of export. The indiarubber creeper is another important plant, growing wild in the forests. Quite close to the sea the fan palm and coco nut palm also grow. The surface, on the whole, is unsuitable for cattle-rearing, but in clearer districts of the forest-zone cultivation of such tropical products as cotton, indigo, tobacco, cassava, yams, maize, rice, etc., is carried on. The larger wild animals are scarce, but in addition to the chimpanzee there are several small forest-loving animals not found in the savannahs.

2. Inhabitants of the Sudan

The Sudan, as its name implies, is the special home of the typical Negroes, who are marked by projecting jaws, broad, flattened noses, and thick lips. They are found unmixed, however, only in the coast-lands of Guinea, being mingled in the plateau zone with peoples apparently related to the Hamites of North Africa, and derived from a mixture of these with the true Negroes. As regards the people, therefore, no less than the natural features, the Sudan falls naturally into two main divisions, the varying nature of the country having, no doubt, done much to bring about the differences in the races.

The **mixed races** found on the open plateaux of the interior are hardy and energetic, much given to trade, agriculture, and cattle-rearing, and contrast strongly with the generally passive and superstitious natives of the forest-clad lowlands. Thus the interior lands, from being brought into contact with the Northern races, have long been the home of a higher degree of civilisation than the coast-lands, even the Negroes having to a large extent adopted the religion—and with it a certain capacity for political organisation—of the invaders from the north. The most important of these mixed races are the Fulbe or Fellata, who, originally cattle-rearers in the Futa Jallon highlands, have now almost everywhere become the masters of the rest of the inhabitants, and the Hausas, a race of born traders dwelling between the Niger and Lake Chad.

Other races of the plateau are more like the Negro in type. The Wolofs, intensely black, dwell between the Senegal and Gambia. The Mandingos, who formerly founded the powerful empire of Melli, are widely spread along the southern border of the plateau, especially near the Upper Niger. They are now mostly subject to the Fulbe. The Songhay, likewise historically important, dwell on the Middle Niger, east of Timbuktu, on the borders of the Sahara. The Kanuri, in the Lake Chad basin, are also of Negro-Hamitic type. Within the bend of the Niger the Negro states of Mossi (inhabited by the Mo or Mohr—capital Wagadugu) and Borgu (inhabited by the Bariba—capital Nikki) have so far maintained their independence, apart from northern influences, and, according to Monteil, a certain degree of civilisation has been attained in Mossi. The Gurunga, on the Upper Volta, are likewise Negroes and pagans.

The pure **Negroes**, inhabiting the coast-lands, are split up into a variety of tribes speaking different languages. Among those speaking the "Chi" and "Ewe" languages respectively, the states of Ashanti and Dahome arose, which have only lately fallen under European influence. They have been the home of debasing superstitions, and human sacrifices have been a regular institution down to the present day. The Yorubas are, however, a superior race, possibly owing to the more favourable nature of their country, which extends over the comparatively open plains west of the Lower Niger. With Ilorin as a capital they once formed a powerful empire.

The Kru, who inhabit the coast near Cape Palmas, are of Mandingan stock, and are a powerful and energetic race, largely employed as labourers ("Kru-boys") along the coast and for the crews of European trading vessels.

Occupations. Trade.—As a general rule the Negro inhabitants of the coast-lands are agriculturalists, growing the banana in large quantities, besides corn, while in the interior both agriculture and cattle-rearing are practised. The dominant Fulbe are herdsmen, and outside their domain it is usually Arabs, or those nearest akin to them, who are engaged in this way, while the negroid peoples are cultivators and traders. The Hausas are the great trading people of the Sudan, and their language is the general medium of intercourse from the Senegal to the Shari, as well as in parts of the Sahara. They are also largely recruited as police by the European powers on the

Gulf of Guinea. In the Negro zone the Yorubas are active traders. The principal long-established trade routes within the Sudan pass from east to west, connecting the various countries in the plateau zone, instead of from north to south. The reason for this is the difficulty of travelling through the forest zone, the rivers being often the only feasible routes, and few of these are really important highways. Even the Niger is only now becoming, in European hands, an important way of approach to the interior, as its swampy forest-clad delta long hindered its use. The great native trading centres lie either near the border of the Sahara at the termini of the desert routes, or near the southern edge of the plateau, just outside the forest zone of the coast. The latter group is represented by such places as Kong and Bontuku in the Mandingan region, Salaga and Yendi in the Volta basin, and Ilorin, the old Yoruba capital. These, though not far removed from the coast, have hitherto been in closer relation with the countries of the plateau; but now that European activity in the Gulf of Guinea is increasing, they are likely to find an outlet rather in that direction.

3. *Native States*

Although almost all the native states of the Sudan fall within the spheres of one or other of the European Powers, their internal administration still remains under the native rulers, so that they may be briefly spoken of before we consider the European possessions.

Sokoto and Dependencies.—*Sokoto*, occupying a great part of the space between the Niger and Lake Chad, is the modern representative of the Fulbe kingdom, established early in the nineteenth century in the domain of the Hausas, who now form the subject population. After the death of the Fulah leader, Othman, the western part of the kingdom was formed into the independent state of Massina on the Upper Niger, while Gando, on the Middle Niger, remained subordinate, but under a different ruler. The city of Sokoto has been replaced as the capital by Wurnu, twenty miles higher up the river of Sokoto. Many of the Hausa towns are almost independent of the central authority, only paying a yearly tribute. Kano, south-east of Sokoto, is the commercial capital of the central Sudan, being the terminus of one of the principal routes across the Sahara. It manufactures blue cotton cloth (exported to many parts of the Sahara and Sudan) and leather goods, and its

Fig. 16.—Kano. (After Barth.)

wares include besides, gold dust, ivory, slaves, salt, natron, earthenware, and many other commodities. At the season when the caravans arrive from distant parts its streets are thronged with merchants, and present a lively spectacle. The population at these times reaches a total of 60,000. Other important centres are Katsena, north-west of Kano, which it once rivalled as a trade centre; Zaria, south-west of Kano; and Yakoba (said to be even larger than Kano), Gombe, Muri, and Keffi, in the more southern parts of the kingdom.

The state of *Gando* stretches along the Middle Niger, and is supposed to exercise some authority over the district of Gurma, west of the Niger towards the north, and that of Nupe, on both banks of the river above its confluence with the Benue.

Adamawa, occupying both banks of the Upper Benue, and stretching over parts of the plateau to the south, is also nominally dependent on Sokoto. Its surface is very broken, and it is split up into a number of minor states, some in the hands of the Fulbe invaders, some still Negro. The chief town and centre of trade is Yola, on the Benue, and Garua, higher up the stream, is also important. Slaves, obtained by raids to the south, have long been exported, and much ivory is collected from the regions south and east by the Hausas, who here, too, have the chief trade and industries in their hands.

Bornu, inhabited principally by the Kanuri, a Negroid people, unattractive both in feature and mental characteristics, dates back as a kingdom as far as the ninth century A.D., when the Mohammedan religion was already introduced. The Fulbe have not obtained the upper hand there. The capital, Kuka, a walled town of 60,000 inhabitants, has long been an important trade-centre as terminus of the Bilma route across the Sahara, but both this trade and the kingdom have lately declined, and the country has fallen a prey to the invasion of the chief Rabah, who was formerly a slave in the Egyptian Sudan. The country shows some degree of barbaric civilisation, an army of 30,000 men having been kept up by the Sultan. The soil is fertile and produces abundant crops. The Bornu breed of horses is also famous. Slaves, ivory, and ostrich feathers have been hitherto the principal exports, and cottons and other European goods, sugar, etc., have been imported by way of the desert. A profitable trade might, no doubt, be opened with the Lower Niger.

Baghirmi occupies the plains watered by the Lower Shari and its tributaries. Its importance has declined of late, as it has fallen largely under the influence of Wadai. Quite recently it has been overrun by the armed bands of the ex-slave Rabah. The population (estimated at a little over 1,000,000) is pure Negro to the south, but much mixed to the north. The dominant race is said to have come from Sennar, on the Nile, in the sixteenth century, when the present capital, Masenya, was founded. Millet and Sorghum, cotton and indigo, are grown, and cattle are reared principally by the Shua, an Arab tribe.

Wadai, lying east of Lake Chad, and including the mountainous district of Borku, in the Sahara, and Dar Runga, in the south, is the only state of the Sudan not yet included in any European sphere. During the reign of Sheikh Ali, father of the present king, it much increased in importance, and in 1871 Baghirmi was conquered for the second time in the century. Kanem, a pastoral district north-east of Lake Chad, formerly itself an important state, has long been disputed ground between Bornu and Wadai. The Mabas, a Negro race professing the Mohammedan religion, are the rulers in Wadai, but many Fulbe and Arabs have come into the country and introduced their dress and customs. The capital is Abeshe, which in 1863 took the place of Wara, now in ruins. The population (about 2,500,000) is sparse, and the people are uncultured and exclusive. The soil is generally poor, but many cattle are reared. Ivory and ostrich feathers are exported *via* the Kufra oasis to the Mediterranean coast, another route leading to the Nile being closed by the Mahdi's rebellion.

Liberia, occupying about 400 miles of the Guinea coast, mostly west of Cape Palmas, has already been alluded to (p. 52). Its capital is Monrovia, at the mouth of the St. Paul River. Coffee, of a variety which takes its name from the country, is an important product. Most of the trade is with Great Britain.

1. *British Possessions*

As most of the European possessions in the Western Sudan originated in the possession of trading stations on the coast, which repeatedly changed hands, and for many years were not accompanied by any territorial acquisitions, they are much split up and intermingled. Thus the British colonies and protectorates occupy four separate parts of the coast, forming six separate administrative areas. We will take these in order from west to east.

The Gambia.—As early as 1618 a fort was erected on the Gambia by an English trading company. After the abolition of slavery the settlement was abandoned, and only renewed (at Bathurst) in 1816 by traders forced to leave the Senegal when that river was given up to the French. After being variously connected with other colonies, the Gambia was finally made a separate Crown colony in 1888. The accessibility of the mouth of the river to large vessels, and its navigability for 257 miles to the Barrakunda Fall, gives the settlement some natural advantages, yet its trade remains unimportant and mostly in the hands of the French. Their territory completely surrounds the colony, which is limited to the banks of the river below the rapid. Ground nuts exported to Marseilles form the only important product. Bathurst has a population of 6000, and the whole colony a little over 14,000.

Sierra Leone, though often before visited by British sailors, first became British territory as a settlement of freed slaves in 1787. In 1791-92 a trading company was formed and the numbers of settlers were increased, the colony taking the name of Freetown. The colony was taken over by the Crown in 1807, and in subsequent years its numbers were largely recruited from slaves liberated by British vessels. Other West African colonies have from time to time been attached to it. Like the Gambia, it is now shut in towards the interior by the French possessions, with which its frontier has been laid down by recent agreements. Near the coast the boundary line runs between the Scarcies and Melakori rivers, but the Los Islands, a little to the north, are British. On the south the Manna River divides Sierra Leone from Liberia. The Sierra Leone peninsula, on which Freetown stands, is formed by a range of volcanic mountains. It shelters the estuary of the river of Sierra Leone, so that this forms the best harbour on the coast, and one which has a special importance as the half-way point between England and the Cape.

A trade route of some importance leads into the interior, and steps have already been taken to improve the communications by means of a railway. Yet the French territory will, it is feared, absorb the more distant trade.

The chief exports are the produce of the forest zone, and include palm oil and kernels, kola nuts, rubber, gum-copal, etc. Cotton goods are imported from the United Kingdom. The trade and revenue are increasing, but British influence has penetrated but a little way into the interior, and the coast tribes still act as middlemen in trade. The climate is notoriously unhealthy and the European residents hardly exceed 200.

The Gold Coast was the site of the earliest trading stations of European nations on the coast of Guinea, and British rule has been extended to the whole coast for a distance of 300 miles by the cession from time to time of those belonging to other Powers. The earliest British settlement was made in 1618. Although British authority has been extended inland during the nineteenth century, and since the military operations of 1896 definitely embraces the kingdom of Ashanti, the actual settlements are still confined to the coast, where, after passing from the hands of a company to the Crown and back again, they now form a Crown colony. The principal towns are Cape Coast (Cabo Corso) Castle (10,000), which, after belonging to the Portuguese and Dutch, passed to the English in 1665, and Accra (20,000), which has of late years taken the place of Cape Coast as the capital of the colony. Other settlements are Axim and Elmina, west, and Addah and Quittah, east, of Cape Coast Castle. There is not one good harbour on the coast, a fact which has been a great obstacle to the development of trade, and the only good road into the interior is that starting from Cape Coast in the direction of Kumasi, capital of Ashanti. British influence is not, as yet, acknowledged by France farther than 9° N., so that trade with the far interior may be limited, as in the case of Sierra Leone, but the coast strip here is unusually rich both in vegetable and mineral products, and the diversity of the surface is another advantage. The abundance of gold (mostly in the form of gold dust) has given the coast its name, and an increasing amount is exported, but in value palm oil and rubber take the first place. Much valuable timber is also produced. Besides cotton goods, a large amount of rum is imported, despite the protests of philanthropists.

Lagos was ceded to Great Britain by its native king in 1861, and its territory was enlarged in 1885 by the

FIG. 17.—CAPE COAST CASTLE.
(Photograph by Rev. J. T. F. Halligey.)

inclusion of the Mahin district adjoining the Niger delta, as far as 5° 10′ E. As the chief outlet for the Yoruba countries, it had acquired some importance in the days of the slave trade, and since coming into British hands legitimate trade has made great strides, being now considerably larger than that of the whole Gold Coast. The actual British territory does not extend more than 30 miles from the coast, but by treaties with the Yorubas in 1891-93 the whole of their country is virtually under British protection, and the limitations placed on trade by their desire to keep it in their own hands have been at last removed. The Yorubas are grouped into a number of large cities, walled as a protection from slave-raiders. Among the largest are Abeokuta (100,000?), Ibadan (150,000), and Oyo (about 70,000). Lagos (75,000) stands on an island, forming part of the belt of sand separating the coast lagoons from the sea. By far the greatest part of the trade is that in palm oil and palm kernels.

Niger Coast Protectorate.—The Niger delta and adjoining districts to the east as far as the Rio del Rey were first definitely placed under British protection in 1884, although British traders had long been active on the various streams known as the "Oil Rivers." This district has remained separate from the territories of the Royal Niger Company (to which, however, the country on each side of the Nun branch of the Niger has been assigned), and in 1891 was placed under an imperial administrator. The trade, chiefly in palm oil, and with it the revenue, has largely increased since that year.

The Niger Company's territories include all the lands on the lower river not included in the protectorate, and stretch up to the Lagos frontier on the west, and the German territory of Cameroons on the east. Treaties have also been made with the rulers of Sokoto, Gando, Borgu, and Yola, securing the influence of the Company in their territories, while by agreements with France and Germany the greater part of Bornu, in addition, has been recognised as within the British sphere. The northern boundaries, however, still require precise definition. Some account has been given of the founding of the Company at p. 57. During its operations a complete administrative system with courts of justice has been introduced, and a military force, chiefly Hausas, under European officers, is maintained. Trading stations have been founded at various points on the river, which, with its tributary, the Benue, affords a good navigable route into the Central Sudan[1] in spite of the disadvantages of the bars at its mouths and the unhealthiness of the low-lying delta. When this is left behind the country improves, and near

---

[1] The Niger is first broken by rapids a little above Rabba. The Benue is free from obstruction for a longer distance.

the confluence of the two streams bare hills replace the forests of the lowlands. The chief settlements are Akasa, at the Nun mouth of the river; Asaba, the capital, just above the delta, with the mission-station, Onitsha, opposite; Lokoja, opposite the Benue confluence; Egga, on the south bank, and Rabba, on the north, in the province of Nupe (subject to Gando); and Ibi and Yola, on the south bank of the Benue. The Niger River is the natural outlet for the populous states of the Central Sudan, which, with their chief centres, have been already described, and there is every probability that a large trade will in time spring up.

Both on the Niger and in the other West African colonies, missionaries have been at work for many years, but the native superstitions are very hard to eradicate.

5. *French Possessions*

By securing the greater part of the interior of the Western Sudan, France has united almost all her possessions into one continuous dominion, which bars the advance inland of the British and other colonies on the coasts. Towards the north this vast territory is continuous with the French sphere of influence in the Sahara, and is thus connected through it with Algeria. The eastern limits are not completely defined. The bulk of the French territory is divided officially into four parts, separate as regards their internal affairs, but all under one Governor-General. The territory on the Slave Coast is an independent government.

(1) **Senegal** embraces all the region of the Lower Senegal and Gambia rivers, except the British colony of the Gambia, which it surrounds completely, including the basin of the Casamanza to the south. It is the oldest French possession in West Africa, and has served as the base whence an advance has been made inland. The French began to search for a way towards Timbuktu as early as the seventeenth century, during which St. Louis was founded; but it was only in the nineteenth century, and especially after 1880, that any great advance was made.

Since then military campaigns have been carried out, by which the powerful Mohammedan chiefs, Ahmadu and Samory, have been overcome; and finally, in 1894, Timbuktu itself was occupied. St. Louis, the capital (20,000), is placed on an island at the mouth of the Senegal. Owing to the shifting bar of the river it can never become an important port. It is joined by a railway with Dakar, which lies in a sheltered bay south of Cape Verde, and has many advantages as a port; yet its trade increases but slowly. Another port is Rufisque, on the same bay, in which also lies the islet of Gorée, strongly fortified. The principal exports are ground nuts and other vegetable products, including rubber, gums, and some timber. The total value, three-fourths of which go to France, is lower than that from the single British colony of Lagos.

(2) **French Sudan** includes the whole region of the Upper Senegal, Gambia, and Niger rivers, with an indefinite extension down the Niger past Timbuktu. The administrative centre is Kayes, on the Senegal. Other important posts are Bamako, where the route from the Senegal strikes the Niger; Segu, lower down the Niger, lately Ahmadu's capital; Jenne, still lower, formerly capital of Massina; and finally Timbuktu, a few miles north of the Niger, but connected at high water by a winding channel with its port, Kabara. The position of Timbuktu, on the margin of the desert, just where the Niger has almost reached its most northerly point, makes it a great centre of exchange for the products of the north and south, and two desert routes converge on it from the north; but it has much declined from its former importance. The state of Massina, which once included Timbuktu, is held by a native ruler under French protection. Its capital is now Bandiagara, south of the Niger. In the south-west and south the French are now supreme in the highlands of Futa Jallon, in Samory's former kingdom (Wassulu), with its capital Bissandugu, and in the important trade centre of Kong, farther east.

Not much has been done to develop the trade of the regions occupied, which remain under military rule. A railway, starting from Kayes on the Senegal, was begun with the intention of connecting that river with the Niger, but it has not advanced beyond Bafulabe, on the Upper Senegal. A telegraph line has, however, been continued to the Niger.

(3) **French Guinea, or Rivières du Sud**, united to Senegal until 1890, includes the coast between Portuguese Guinea and Sierra Leone, and borders on Futa Jallon inland. It is under a lieutenant-governor, who resides at Konakri, on a small island near the mouth of the Dubreka River. Other posts are Boke and Victoria, on the Nuñez River; Boffa, on the Pongo; and Benti and Melakori, on the Melakori River.

(4) The **Ivory Coast** lies between Liberia and the British Gold Coast. The capital, Grand Bassam, is situated, like Lagos, on the narrow belt of sand which separates the coast lagoons from the sea. It was first occupied by the French in 1842, but has come into importance only since Captain Binger's journey of 1891-92, which showed that the rivers of this district come from farther inland than had been supposed, the Komoe and Bandama supplying routes through the forest belt towards the important town of Kong.

Dahome.—The French, English, and Dutch all had forts at Whyda, on the Slave Coast, in the eighteenth century in the days of the slave trade, but no permanent acquisition was made until in 1884-85 France took possession of the posts of Great Popo and Porto Novo. French influence has lately been extended inland by the conquest of the Negro kingdom of Dahome (capital, Abome). Palm oil and some gold dust are exported.

6. *Other European Possessions*

Portuguese Guinea.—The only part of the former extensive Portuguese possessions in Guinea which is retained at the present day is a small section of the coast between the French provinces of Senegal and Rivières du Sud, with the Bissagós Islands lying off the mainland.

This territory stretches only about 150 miles inland, being completely enclosed by the French Sudan. The coast is broken by the broad estuaries of the Cacheo, Geba, and Rio Grande. Bissao, on an island at the mouth of the Geba, is the principal station, but the trade is insignificant.

The **German** territory of **Togoland** occupies a small part of the Slave Coast between the British Gold Coast colony and the French districts on the Bight of Benin. Togo, which gives the territory its name, lies on the landward side of the coast lagoon, while between this and the sea are the towns of Bagida and Little Popo. All these were annexed in 1884, and much has been done since to extend German influence into the interior. As the boundary with the Gold Coast is formed in great part by the course of the Volta, Togoland widens inland, stretching behind the eastern part of the Gold Coast territory. North of about 8° 12′ N. a neutral zone has been formed between the two colonies, containing the trade centres of Salaga and Yendi. On the more or less healthy uplands of the interior of Togoland the stations of Misahöhe and Bismarckburg have been founded. As on the whole Guinea Coast, palm oil and palm kernels are the principal exports.

CHAPTER VIII

THE NILE REGION

1. *Physical Features*

The last three chapters have had to do with three different zones of country (the northern coast-lands, the desert, and the Sudan), differing both in climate and productions, which together make up the greater part of the northern half of Africa. But so far we have left out of consideration the eastern parts of the separate regions, which are bound together, as it were, into a whole by the course of the Nile. Passing through each of the three in turn, the river brings them into relation with each other, so that both on natural and historical grounds the Nile region may be treated as one. The head-waters of the Nile lie on the high plateau of Southern Africa, and belong to a region distinct in many ways, which will be dealt with in a later chapter. They are therefore only referred to here so far as is necessary to give a connected idea of the Nile system as a whole. The rest of the Nile valley, which forms the central line of the Nile region, consists of a shallow trough or hollow running almost due north, which has been scooped out by the action of running water in the North African plateau. It is bounded on the east by highlands throughout its whole extent, but on the west the desert which borders the valley during its second half has only a moderate height.

The Nile.—The sources of the principal stream, which were for many centuries involved in mystery, are formed by three lakes lying on the high East African plateau, with the various streams which flow into them. The Victoria Nyanza, to the east, is the largest lake in Africa, and is situated on the equator, at an altitude of 3900 feet. It is a vast, nearly circular, basin, 200 miles across, into which rivers run from all sides. The principal stream is the Kagera, which comes from the south-west. As a rule the shores of the lake slope gently, or at the most are of the form of low cliffs. About the middle of the northern shore the Nile (here known as the Somerset Nile) leaves the lake, falling over a rocky ledge by the Ripon Falls. It then takes a north-west course, and rushes down from a higher to a lower level by a succession of rapids, which end in the picturesque Murchison Falls (120 feet high). The two western lakes lie north and south in a narrow trough, bounded in great part by high and steep mountain walls, which rise in the interval between them into a range of snowy summits on the east. The southern of the two, named Albert Edward, lies at a height of 3200 feet above the sea. It is of very irregular shape, consisting of an oval basin to the west united by a narrow channel to a long narrow arm running north-east. At the western end of the north shore a river issues, known finally as the Semliki, which, after passing through dense forests and receiving copious supplies of water from the snowy peaks of Ruwenzori, flows into the northern lake, the Albert Nyanza (2300 feet), across a bare, level plain, which must formerly have been covered by its waters, but which has in time been raised above its surface by the deposits of the river. The Albert Nyanza is even now shrinking in size, and on the west side a flat shore stretches along the foot of the steep cliffs by which it is bounded. Near its north end it receives the Somerset Nile, just after it has precipitated itself down the Murchison Falls.

The united stream of the Nile leaves the Albert Nyanza at its northern point, and is again broken by rapids for some distance. But after passing the fifth parallel of north latitude it reaches a level plain, where its current becomes slow, and is choked at times by great masses of floating vegetation. A little before 10 N.

it receives from the west the important tributary, the Bahr-el Ghazal, which brings in the combined waters of a wide extent of country to the west. It is the last permanent tributary from this direction. The last part of its course is swampy and choked with vegetation like the main river. After this the Nile makes a sudden bend

FIG. 18.—THE UPPER NILE.
(After Junker.)

to the east, and before returning to its northerly direction receives from the south-east the Sobat, a very winding river, whose sources have not yet been explored. After passing 15° N. latitude, this southern branch of the Nile, known as the White River or Bahr-el-Abiad, is joined by the great eastern branch, the Bahr-el-Azrek or Blue River, and after 180 miles more by the Atbara, both having their sources in the highlands of Abyssinia. The Atbara is the last permanent stream which joins the Nile, which here begins to flow across the desert, describing a great S-shaped curve, which adds much to the length of the stream. Owing to the intense evaporation, it becomes less and less in volume during the 1500 miles which have to be passed over before it reaches the Mediterranean Sea. During this distance it for the most part flows through a narrow valley bounded by lines of hills.

Between the junction of the Blue and White rivers and its mouth the Nile is broken at six places by cataracts, which are known as the first, second, third, and so on, beginning from the one nearest the sea. The sixth occurs between the Blue Nile and the Atbara. These cataracts are not single waterfalls, but are made up of a succession of rapids caused by rocks in the bed of the stream. They do not entirely prevent navigation at all seasons of the year. At some of them the river is broken up at low water by innumerable jagged ridges of rock, between which the water flows in narrow channels. On reaching the sea the Nile forms a delta composed of fertile alluvial soil thrown down by the river. Of the various branches into which it is split up the two principal are those which enter the Mediterranean at Rosetta and Damietta, from which places they take their names.

Its Inundations.—The great importance of the Nile to the countries through which it flows lies in its yearly inundations. Most tropical rivers have an annual period of flood, but in the case of the Nile this is unusually well marked, and it is of vital moment to the dwellers on its banks. Were it not for the river, with its yearly overflow, the region traversed by the Nile during the latter part of its course would be as barren as the rest of the desert, but as it is, the river supplies not only moisture, but also a most fertile layer of alluvial soil. The cause of this annual overflow of the Nile seemed mysterious to the ancients, as the country from which it came retained the same desert character, as far as they knew it. It has only been clearly explained by the results of modern explorations. Both the Blue and the White rivers play their part in promoting the inundation. The three great lakes from which the White River takes its rise act as reservoirs for the abundant equatorial rainfall, so that this branch has a constant supply of water through the year. The Blue River is fed by torrential rains, which fall periodically in the Abyssinian highlands, so that the channel of the river, which remains at a low level during the remainder of the year, is suddenly filled, and carries down a great body of water, charged with fertilising sediment, to meet the steady flow of the White River. The result is the inundation of the banks of the Nile in the lower part of its course, where the whole flat valley as far as the bounding hills becomes in time entirely covered. The rise of the Nile occurs with such regularity that in a normal year its various stages can be predicted within a very few days. It begins in June, and the flood is at its height in September and the first part of October.

The Upper Nile Country.—The countries on the Upper Nile resemble in some ways the zone in West Africa which touches the desert on the south, and may on this account be termed the Nilotic or Eastern Sudan. The same gradual transition is to be seen from a luxuriant and rainy region in the south to one dryer and more bare as the borders of the desert are approached. As in West Africa, this region is inhabited by the branch of the Negro race which has been termed Sudanese. Although watered by tributaries of the Nile, the highlands to the east are separated, both naturally and politically, from the Nile countries, and may therefore be reserved for another chapter.

Apart from these, the Nilotic Sudan may be divided into a northern and a southern section, at about 12° N. latitude, or a little below the mouth of the Bahr-el-Ghazal. South of this the country belongs to the region of savannahs and forests. The larger part lies to the west of the Nile, where the land watered by the Bahr-el-Ghazal and its affluents, with one or two direct tributaries of the Nile, forms a well-defined region in itself. The surface is mainly undulating, rising gradually to the south-west towards the Nile-Congo watershed, which has rather the character of a broad and gradual swelling of the surface than a mountain range, and is furrowed by numberless streams flowing in more or less parallel courses. The beds of these streams are mostly steep-sided troughs cut down below the surface of the plain and filled with a thick growth of vegetation, forming a sort of river avenue. Occasional isolated ridges or groups of hills rise above the general level, and in the south-east reach a height of over 6500 feet. Here the affluents of the Welle, belonging to the Congo system, approach very closely to the Nile, their sources lying only about 30 miles distant.

North of about 12° the country is dryer and has more of a steppe character, with a thorny scrub vegetation. Except the Abyssinian tributaries of the Nile, the watercourses are only periodically filled. West of the Nile the average elevation of the country is higher than that of the Bahr-el-Ghazal region, being occupied by a projecting arm of the high Central African plateau mostly over 1500 feet above the sea, and rising still higher towards the west, where the rugged mountains of Darfur occupy a large area. The highest ridge, Jebel Marrah, is about 6000 feet high. It forms the water-parting between the periodical streams flowing east and south-east to the Nile, and those flowing west and south-west towards Lake Chad.

The Nile Deserts.—Even before receiving the Atbara, its last tributary, the Nile has entered the desert zone, through which it flows uninterruptedly to its delta. The immediate valley of the river is in many ways similar to the oases scattered elsewhere over the desert, the date palm being the characteristic plant. On the west the desert is ushered in by the Bahiuda steppe, not absolutely barren, but watered by occasional torrents, and covered with poor scrub in parts. North of this lies the vast region of shifting sands known as the Libyan desert, in great measure untrodden by man. Except quite to the south, where the Wadi Melk runs towards the Nile from the Darfur highlands, even intermittent water-courses are absent, and, so far as is known, the monotony of the surface is broken by no important ranges of hills. The few scattered oases are found chiefly towards the north, where they occur in two groups. The more northern stretches east and west, and is the continuation of the line of oases in Tripoli south of the coast plateau, which is likewise continued eastwards towards the Nile. The principal oasis, Siwah, was known to the ancients as the oasis of Jupiter Ammon. All the oases in this line are below the level of the sea. The more southern group runs more or less parallel to the Nile, beginning with the "Little" or Bahrieh Oasis, and containing besides those of Farafrah, Dakhel, and Khargheh (the last being also known as the "Great" Oasis).

On the east of the Nile the plateau attains a much greater height than on the west. The Nubian desert, which is about as inhospitable as the Libyan, lies within the northern limb of the great S-shaped bend of the Nile. It is somewhat broken by rocky ridges and dry watercourses. As the Nile again turns eastwards after its great westerly bend, the space between it and the Red Sea is much diminished, and only a narrow strip of desert intervenes between the river and the range of hills which bounds the plateau on the east, and with few interruptions skirts the Red Sea from the north end of the Abyssinian highlands to the head of the Gulf of Suez. This range maintains a considerable height throughout, although broken up at intervals into isolated peaks and ridges, with passes between them leading from the Red Sea to the Nile. The principal mountain groups are the Jebel Elba, with its highest peak, the Jebel Soturba (6900 feet), in about 22° N.; the Jebel Hamada (over 6000 feet) and Jebel Zebara (7500 feet), a little south of 25°; the Jebel Um Delpha (7150 feet) and the Jebel Dukhan, a little south, and the Jebel Gharib (7880 ? feet), a little north, of the mouth of the Gulf of Suez.

Although mostly barren at the present day, these hills are seamed with ravines which must have been hollowed out when the rainfall was much greater than it is now. Sudden storms still occasionally occur, causing torrents of water to rush down the Wadis, and snow is not unknown on the summits. Some of the valleys in which the moisture collects possess a fair amount of vegetation.

I

Several bold promontories run down from the seaward face of this range. No permanent streams drain its slopes, but at the south extremity the Khor Baraka brings down during the rains a considerable volume of water from the Abyssinian highlands, which its valley separates from the more northern hills.

The Fayum and Delta. — At about 500 miles from its mouth the Nile makes a sharp bend to the west and north-west, and soon after this an artificial branch diverges to the left and flows parallel with the main stream for some hundreds of miles. It is known as the Bahr Yusuf, or River of Joseph. Shortly before reaching the Delta it sends an arm westwards, through a gap in the limestone plateau, into the remarkable depression of the Fayum (in about 29° 20′ N. lat.). Though once filled with water, the Fayum has mostly been reclaimed, and is one of the most fertile districts of Egypt, being irrigated during high Nile by channels derived from the Bahr Yusuf. The surplus waters find their way into a small remnant of the old lake called Birket-el-Kerun, 140 feet below the level of the sea. A little south of the Fayum, but separated from it by higher ground, there is a second depression—the Raian—also below sea-level, which it has been proposed to utilise as a storage reservoir of the Nile flood-water.

Although the Fayum has a somewhat deltaic character, the Delta proper does not begin until after the Nile has crossed 30° N. latitude. Shortly after this the river splits into its two main branches, the only ones, in fact, which now reach the sea. The western or Rosetta branch bends somewhat to the west, keeping at first close to the edge of the western plateau, but afterwards, as this curves away from the river, the deltaic land extends some distance to the west of the left bank. On the east the high ground ceases almost at once, and a large part of the Delta lies on this side of the eastern or Damietta branch, merging at last into the level sands which form the greater part of the Isthmus of Suez. The whole of the level alluvial land is traversed by sluggish channels, the remnants of former Nile mouths, which now empty themselves into a series of shallow lakes or lagoons, cut off from the sea by narrow banks of sand. These lagoons extend round the face of the Delta with hardly an interruption, except where the two main branches have created firm land by depositing their sediment. The largest are the Lake of Burlos, between the two main mouths, and that of Menzaleh, east of the Damietta mouth.

The irrigation of the Delta is regulated by a great weir called the Barrage, built across the stream near the point where the two branches separate. This prevents the water from flowing away too quickly to the sea, and allows it to be distributed by canals over the whole Delta, so that a supply is everywhere available for the crops.

Climate, Flora, and Fauna.—Stretching over so many degrees of latitude, the Nile region shows great differences of climate between its separate sections. These differences consist rather in variations of rainfall than of temperature, for, except near the shores of the Mediterranean, a generally high temperature prevails over the whole region. The southern section of the Nilotic Sudan has a copious rainfall spread over a large part of the year, but broadly divided into two rainy seasons, with a short dry period between. Farther north the rainfall diminishes and the greater part of the year is dry, and still farther north the Nile passes through the almost rainless zone, which continues almost to its mouth, and is subject (in summer) to the most intense heat of all. The climate of the Delta is marked by comparative coolness during the winter months, in which north winds prevail, bringing fogs from the sea, with some rain. The country may be said to be most healthy where the rainfall is least, the southern parts of the Sudan coming within the unhealthy malarial region of Central Africa, while the rainless zone, in spite of its heat, is healthy like the rest of the desert. Lower Egypt also is healthy, except during the subsidence of the annual floods, when fever and dysentery prevail.

The plants and animals of the Nile region vary with the climate, the rich tropical vegetation of the rainy zone ceasing entirely in the northern half of the region, which is marked by an almost total absence of trees. In Egypt every inch of fertile soil is taken up for cultivation, and, except the date palm and some of the fruit-trees of the Mediterranean region, almost the only tree is the sycamore fig. The banks of the Nile are often fringed with papyrus, especially in the marshy section of the river. Like the lotus, for which Egypt was once famous, the papyrus is almost exterminated in the lower section of the river.

As regards the animals, the Nile countries fall within two quite distinct regions. In the south such forms as the elephant, rhinoceros, hippopotamus, giraffe, ostrich, and crocodile, as well as various antelopes, are the most striking, and in the south-west the chimpanzee, one of the man-like apes, occurs. In Egypt proper all these are wanting, the crocodile and hippopotamus, which used to be seen in the waters of the Lower Nile, having retired farther south, and the leopard, hyaena, fox, and jackal are the only representatives of the larger wild animals. Many kinds of birds are, however, to be seen, including vultures, the stork, pelican, flamingo, ibis, and other water birds, together with some winter migrants from Europe. The camel is in use as a domestic animal in all but the southern parts of the Sudan.

2. *Inhabitants of the Nile Region*

The line of division between the lighter northern races and the black or Negro race of the south can be drawn with still less precision in the Nile region than in West Africa. In the west the term Sudan (country of the Blacks) is justified by the fact that in the whole of the zone so named the great bulk of the people belong to the black race. But on the Nile, perhaps because of the natural facilities for southward movement afforded by the valley of the river, we find that the northern races have far overstepped the southern limit of the desert, and in that part to which the term "Egyptian Sudan" has been especially applied — that lying around the junction of the Blue and White rivers — the larger part of the population is of Arab descent. The Negroes are found as the bulk of the population only in the southern of the two zones into which we have divided the Nilotic Sudan, *i.e.* broadly speaking, south of 12° N. latitude. North of this, however, they occur in sufficient numbers to allow the country to appear to the dwellers in the north as the land of the blacks, and they must once have formed a larger proportion of the population than at present. Even in the desert zone the Nubians show signs of a strong admixture of Negro blood, while the Fur and Nuba in Darfur and Kordofan are regarded as true Negroes.

The Hamites and Semites.—The distribution of these two races is almost as much confused in the Nile region as elsewhere. The Hamites form in Egypt the bulk of the *fellahin* or lower orders, the descendants of the ancient Egyptians, occupied chiefly in agriculture. Farther south, under the name Bisharin or Beja, they occupy most of the area between the Nile and the Red Sea as far as Abyssinia, one of the most important tribes being that of the Hadendoas. The Nubians, too, are supposed to have some strain of Hamite as well as Negro and Semitic blood. The pastoral Bedawin of Egypt, on the other hand, are Semites, and the whole of the country west of the Nile is inhabited chiefly by Arab tribes down to about 12° N. latitude, including parts of Kordofan and Darfur. The most important tribes are those of the Kababish, who inhabit the desert to the south and west of the great Nile bend, and the Bagara, between Darfur and the Nile. The Arabs and Nubians have been most active in pushing southwards up the Nile valley, and from their ranks have issued the slave-raiders who, with Khartum as a base, have had such a baneful influence on the Negro countries beyond. They also have been the chief supporters of the military revolt under the fanatical leaders, the Mahdi and Khalifa.

The Negroes of the Nile.—South of about 12° N. the region of the Nile and of its western affluents is occupied by a number of populous Negro tribes who show no admixture of northern blood, the inroads of Arabs and others from the north dating from comparatively recent years. Two main groups may be distinguished.

The first includes what have been termed the Swamp Negroes, dwelling on the swampy banks of the Nile and the lower courses of the Bahr-el-Ghazal and other tributaries. The Shilluk and Dinka are the most important tribes. The Dinka are especially noted as cattle-rearers, in which they stand apart from the great bulk of the Sudan Negroes, who are typically agriculturalists. The second group includes the Bongo and other tribes dwelling on the upper courses of the western streams. These are marked by a reddish tinge of skin, and are not cattle-rearers, but are all accustomed to iron-working, iron being very plentiful in the soil of their country. Beyond the Nile-Congo water-parting, in a region little distinguished from the Bahr-el-Ghazal region by its physical features, there are other tribes bordering on the Bantu populations of the south, who are generally classed with the Sudanese Negroes, although they have doubtless become much mixed with their Bantu neighbours. These are the A-Zandé (also called Niam-Niam), noted for their cannibalism, and the Mangbattu, also cannibals, but possessed of a comparatively high social organisation.

3. *Egypt*

No other country in the world has a history dating so far back as that of Egypt, the home of the earliest civilisation of which we have any record, and the country where the art of alphabetical writing was first invented. At a time when the greater part of the world was peopled by savage tribes the old Egyptians were already skilled in agriculture, in architecture, and other arts, and even in the science of astronomy, as is shown by the way in which the great pyramid of Gizeh is set with its sides exactly facing the four cardinal points of the compass. In the thousands of years which have elapsed since those early days, the country has passed through many vicissitudes, remaining for centuries under the rule of a succession of native dynasties, and then falling under the dominion of one foreign nation after another. Persians, Greeks, Romans, Arabs, and Turks have in turn held sway, and

Egypt still remains a subject nation, the ultimate fate of which engages the diplomatic consideration of several of the great Powers of Europe.

Government, Population, etc.—Though still nominally a part of the Turkish Empire, Egypt has been since the reign of Mehemet Ali (1811-48) virtually independent of the Sultan's control, merely paying a yearly tribute to Turkey. Since 1866 its rulers have borne the title of Khedive. The French invasion under Bonaparte in 1798 —though ultimately unsuccessful—drew the attention of France to the country, and French influence was in the ascendant during the whole reign of Mehemet Ali. It was a Frenchman, too, who both planned and executed the Suez Canal (*see below*). The military revolt of Arabi Pasha in 1882 resulted in the intervention of Great Britain, to which the maintenance of the Suez Canal as an open highway to the East is of paramount importance. A bombardment of the port of Alexandria by British ships, and military operations crowned by the defeat of Arabi's forces at Tel-el-Kebir, placed the country under British control, which is still exercised through officials who act as advisers to the Khedive's Government, supported by an army of occupation. The result of British influence has been a thorough reform of long-standing abuses in the government, a lightening of taxation and abolition of forced labour, an extension of irrigation and other useful public works, and a marked improvement in the revenue and general financial outlook.

The population of Egypt, which may now be held to include a portion of the old province of Nubia, was in 1892 a little short of 7,000,000, the greater part of which is congregated on the alluvial lands of the Delta and the immediate neighbourhood of the Nile, where the annual floods permit the practice of agriculture. These parts of the country are more thickly peopled than any other tract of similar extent in the world, while the rest of the country has very few inhabitants. Egypt has from very early times been divided into a northern and a southern section, known as Lower and Upper Egypt respectively, while under the Turks a third subdivision (Middle Egypt) was formed. Lower Egypt extends southwards along the river to about the position of the Fayum, the boundary, however, bending northwards on the west so as to leave that province on the south. It thus includes only a small part (but that the most fertile and populous) of the Nile valley, but embraces a large area of desert to the west. Upper Egypt has, since 1885, included a portion of the old province of Nubia, which was retained when the rest of the southern provinces were left to the Mahdi; and has been further enlarged by the reconquest, in 1896, of the province of Dongola. Both Upper and Lower Egypt are subdivided into governments, the number of which has varied from time to time, and there are also three governments which embrace the Isthmus of Suez as well as the Sinai Peninsula, properly belonging to Asia.

Chief Towns.—The two largest cities of Egypt and of the whole of Africa are in Lower Egypt. Cairo (375,000) is the seat of Government, and is placed on the Nile just above the point where the various branches divide at the beginning of the Delta; Alexandria (215,000) is the chief port, and lies on the sea-coast almost at the western end of the Delta.

Old Cairo, adjoining the modern city on the south, was founded by the Arab invaders in the seventh century, the capital being moved to its present position in the tenth century. The greater part of the city stands a little away from the Nile, on the lower slopes of the Jebel Mokattam, the last northward spur of the eastern plateau. On the higher slopes to the south-east stands the citadel, which contains the palace, the Mosque of Mehemet Ali, and other public buildings. The city is enclosed by old walls on the north and east, and there are forts on the highest crest of the Mokattam. The suburb of Bulak, to the north-west, on the Nile, is divided from Cairo by gardens, and serves as the river-port of the capital. Its celebrated Museum of Antiquities has lately been removed, first to Gizeh, and afterwards back again to the east side of the Nile. Alexandria, built by Alexander the Great in 332 B.C., and long the chief centre of the commerce and learning of the world, lies on a narrow strip of land be-

tween Lake Mariut (the ancient Mareotis) and the sea.
It has been built out into the sea, so as to join the ancient
island of Pharos (on which is a lighthouse) and separate

Fig. 19.—General View of Cairo.
(Photograph by Lekegian & Co., Cairo.)

the old harbour to the south-west from the new one on
the north-east. By far the greater part of the foreign
trade of Egypt passes through Alexandria, which is joined
by the Mahmudiye Canal to the Rosetta mouth of the Nile.

It is fortified and has a naval arsenal. Its forts were bombarded by a British fleet in 1882 when held by the rebels under Arabi Pasha.

Other maritime towns are Damietta and Rosetta, at the mouths of the two main branches of the Nile. Port Said, at the northern end of the Suez Canal, and Ismailia, nearly midway between its two ends, have sprung up since it was opened, and owe their importance solely to the passage of ships by this route. Suez, at the southern end, has not much benefited by the opening of the canal, but remains a purely Arab town. Suakin, about half-way along the Red Sea coast, has the best harbour on that sea. The town is on an island, in a basin sheltered by reefs of coral, of which material the houses are built. Its position near the easterly bend of the Nile, to which a railway will probably be built some day, gives it the command of one of the best routes to the Egyptian Sudan. Kosseir, farther north, has some importance as a port, owing to a somewhat similar position with respect to the Nile.

The inland towns of Egypt besides Cairo are mostly agricultural centres. In the Delta the chief are Tanta, Damanhur, and Zagazig, and in Upper Egypt, Siut, Keneh, Esneh, Assuan, Korosko, and Wadi Halfa, all on the Nile, and occurring in the above order from north to south. Away from the Nile the chief centres of population are the Fayum and the oases in the Libyan desert.

Communications.—The Nile supplies water communication throughout the whole length of Egypt, broken only by the cataracts during the period of low water, and numerous steamers and the picturesque sailing-boats called dahabiyehs ply both on the branches of the Delta and on the Nile itself. Railways also traverse the Delta in various directions (with a branch to Ismailia and Suez), and a line ascends the Nile on its western bank, passing Assuan, and it will probably be soon carried farther south. A branch line runs westwards to the Fayum. The Nile is spanned near Cairo by a railway bridge, which connects the Delta lines with that up the valley. Caravan routes also lead across Lower Egypt from west to east, connecting the other countries of North Africa with the Nile Delta, and even with Arabia. They are frequented by pilgrims making the journey to Mecca. One of them follows the shore of the Mediterranean, and another the line of oases (Siwah, etc.) just south of 30° N. latitude. Another caravan route leaves the Nile at Siut, and passes through the Great or Khargheh oasis towards Darfur, in the south, but the Mahdist rebellion has practically closed this as a trade route.

Resources and Trade.—The resources of Egypt consist almost entirely in its agricultural products. The alluvial soil deposited by the Nile water is of surpassing fertility, and being renewed year by year, runs no risk of becoming exhausted. It therefore produces many times larger crops than the same area in other countries. It is proposed to further develop the system of irrigation, by which alone this fertility can be fully utilised, by the construction of a great flood reservoir in Upper Egypt above the Island of Philae. The cereals include wheat, maize and barley, durra, and rice, wheat covering the largest area. Lentils and other kinds of pulse are also largely grown. But a crop that has acquired great importance of late years is cotton, which is grown principally in the Delta, as are also maize and rice, while the other crops named, together with sugar, are grown chiefly in Upper Egypt. Cotton is now exported in large quantities, and besides this, wheat and other cereals and sugar make up the greater part of the exports. The imports include cotton and other stuffs, coal, timber, metals, and machinery. The largest share of the trade is in British hands, but Turkey, Austria, France, Italy, and Russia also take their part in it.

The Suez Canal.—The chief modern importance of Egypt lies in the fact that the Suez Canal, which has become the chief route from Europe to the East, passes through its territory. This great work was accomplished in spite of much opposition by the energy and perseverance of the French engineer, M. F. de Lesseps. It was begun in 1859, and opened for traffic in 1869. Its length is about 100 miles, and the minimum depth is now 26 feet, and the minimum bottom width 72 feet. By the use of the electric light, steamers pass through by night as well as day, so that a stream of traffic is always passing to and fro. Owing to the importance of the British possessions and trade in the far East, to which the canal supplies a route shorter by several thousands of miles than the old route *via* the Cape of Good Hope, by far the greater number of steamers passing through the canal are British, amounting to no less than three-quarters of the total tonnage. With the increase of traffic by this canal the Cape route to India (but not that to Australia), has virtually been abandoned except by sailing ships; and as the steamers built for the canal route are by no means suited to brave the storms of the Southern Ocean, communications between England and the East would be completely disorganised in case of a block in the canal. The advantages derived from the canal are thus counterbalanced under existing circumstances by considerable risks.

The canal resembles a great trench dug through the sands of the Isthmus of Suez, in which the fantastic effects of mirage are often to be seen. In the northern section it skirts the eastern shore of the shallow Lagoon or Lake of Menzaleh, belonging to the Nile Delta. The dividing ridge between the Mediterranean and the Red Sea is formed by the narrow plateau of El-Gisr, 50 feet above sea level, south of which the canal passes through Lake Timsah (with the town of Ismailia on its north shore) and the larger Bitter Lakes to the south, which must once have been a part of the Gulf of Suez. Lake Timsah occasionally receives some of the Nile flood-water through the Wadi Tumilat, which is also traversed by a railway and freshwater canal, both of which accompany the course of the ship canal to Suez. The freshwater canal leaves the Nile at Cairo, and furnishes drinking water to the population which has sprung up along the course of the main canal.

Antiquities of Egypt. Everywhere in Egypt there are to be seen the remains of the ancient civilisation of the country, which have been wonderfully preserved owing to the dryness of the atmosphere. The most striking are the huge pyramids, particularly the three situated south-west of Gizeh (opposite Cairo), and associated with

FIG. 20.—THE PYRAMIDS.
(Photograph by Messrs. Lekegian & Co., Cairo.)

the almost equally wonderful Sphinx. Stupendous ruins of ancient cities, with temples, colonnades, and other vast structures, are to be seen on the site of Memphis, a little above Gizeh, on the west bank of the Nile; on that of Thebes, on both banks of the river, a little above the bend at Keneh; and at various other spots along the valley in Upper Egypt and Nubia as far as the so-called Island of Meroe, between the Atbara and the Blue Nile. The rock temples of Ibsambul or Abu-Simbul are especially worthy of mention. The Egyptian monuments include not only the works of the ancient Egyptians, but others dating from the age of the Ptolemies, such as the temple of Horus at Edfu; and from that of the Roman Empire, such as the well-known temple on the Island of Philae, a little above Assuan, built in the reign of Tiberius.

4. *The Nilotic Sudan*

Recent History.—The whole of the region of the Nile within the limits assigned in this chapter was united under the government of Egypt before the rebellion of the Mahdi in 1883 broke off the Sudan provinces. The first advance southwards beyond the limits of Egypt proper was made in the days of Mehemet Ali (1811-48), in whose reign the frontiers were extended southwards, so as to include the "Egyptian Sudan," with its centre at Khartum, founded in 1820-21, at the junction of the Blue and White Niles. This was succeeded by a gradual extension of influence southwards, until in 1870, during the reign of Ismail Pasha, a decided forward move was again made. Under Sir Samuel Baker, and subsequently under General Gordon, Egyptian sovereignty was extended westwards over Darfur and southwards to the shores of the Albert Nyanza, and fortified posts were established both on the Nile itself and in the Bahr-el-Ghazal region, and even beyond the watershed to the south-west. The great obstacle throughout was the opposition of the Arab slave-traders, who saw that they would be unable to enrich themselves as they had done if the European officers had their way.

It was the discontent which prevailed on this account, together with the misrule of the subordinate Egyptian officers, that lent strength to the rising begun in 1881 by the fanatical leader Mohammed Ahmed, who claimed to be the Mahdi or "guided one" destined to reform and renovate the Mohammedan world. The revolt quickly gathered force, and in 1883 an army sent under Hicks Pasha to suppress it was annihilated in Kordofan, and the Mahdi's power soon extended to the very gates of Suakin, on the Red Sea. In 1884 General Gordon accepted a mission for the withdrawal of the Egyptian garrisons from the Sudan, but was soon shut up in Khartum, and reduced to great straits from scarcity of supplies. A British army under Lord Wolseley started up the Nile for the relief of Gordon, but the advanced force reached Khartum in 1885 only to find that the place had been carried by assault two days before and Gordon assassinated. Slatin and Lupton, the governors of Darfur and the Bahr-el-Ghazal province, had already been defeated and taken prisoners; and although Emin still held out in the south until relieved by Stanley in 1889, Egyptian authority was practically confined to the Lower Nile as far as Wadi Halfa, and the actual coast of the Red Sea as far south as Suakin.

Mohammed Ahmed died soon after the fall of Khartum, appointing as his successor the "Khalifa" Abdullahi, a man of the most cruel and tyrannical nature, who has since ruled the subject populations with a rod of iron. His authority extends up the Nile to a little above Lado, and embraces all of the old Egyptian Sudan except the Bahr-el-Ghazal province, where the Dinka Negroes have succeeded in holding their own. A number of European prisoners have remained in the Khalifa's hands, subject to all kinds of indignities, but two of the chief, Father Ohrwalder and Slatin Pasha, have succeeded in escaping, Slatin only in 1895. He had spent no less than eleven years in captivity.

The result of all the troubles to which the Sudan has been subject has been that many districts once prosperous have been reduced to desert wastes, three-quarters of the total population having succumbed. The remainder are mostly little better than slaves, and would, no doubt, at once turn against their present masters if help were given them from outside.

Towns, Resources, etc.—The natural centre of the whole Nilotic Sudan is Khartum. Owing to its important position at the confluence of the Blue and White branches of the Nile, it is the meeting-point of the river routes from the north, south, and south-east, and to it flows also the trade of the whole surrounding region, including Darfur, and even more remote countries in the west. It is built on the tongue of land between the two rivers. Since its capture by the Mahdi, Omdurman, on the opposite bank of the White Nile, has been the centre of the revolted provinces. Other important towns are Berber, just below the Atbara, at the nearest point on the Nile to Suakin; Kasala, now occupied by Italy, at the foot of the Abyssinian highlands; El-Obeid, capital of Kordofan, and El-Fasher, of Darfur; and in the more southern regions, Fashoda, in 10° N., and Lado, in 5° N.

The importance of these southern regions to Egypt lies (apart from their commercial resources) in the fact that they supply the water for the fertilising inundations which alone render agriculture possible, and which might be checked by a hostile power established on the Upper Nile. The chief products are ivory and gums, whilst in Darfur both wheat and tobacco flourish. Of all the former Sudan provinces that of the Bahr-el-Ghazal is said to be the most valuable. It is well watered and has a most fertile soil, producing large quantities of cotton and india-rubber, while elephants are numerous in its mountains and forests. Its population is numerous, and some of its tribes possess cattle in great abundance. Among the resources of the Nilotic Sudan the copper mines of Hofrat-en-Nahas, in the south of Darfur, deserve mention. But that region can only be thrown open once more to civilising influences by the overthrow of the present tyrannical rule of the Khalifa.

CHAPTER IX

NORTH-EAST AFRICA

1. *Physical Features*

General Characters.— In this chapter we have to consider the northern part of the important highlands of East Africa, which lie to the east of the Nile region just described. The northern and southern parts of this high region are not separated by any well-marked natural dividing line, but as a distinction exists both in the inhabitants and the historical and political relations of the two areas, it will be convenient to treat them separately. The small number of large lakes in the northern division supplies a certain natural basis of distinction, and it may be said also that steppe-lands with scrub vegetation are more prevalent in the north, corresponding to a more scanty rainfall. The region we are now considering lies almost in the same latitudes as the Sudan, but has a much greater general altitude, and therefore a cooler climate, and is distinguished especially by the almost complete absence of the Negro race.

The area may be divided roughly into two parts. In the west lie the broad highlands of Abyssinia and the countries south of it, sometimes included under the general term Ethiopia, averaging about 8000 feet above the sea, with their central line running north and south; while a lower region extends to the eastward, in which the chief line of high ground runs east and west, forming, as it were, the backbone of the eastern "horn" of Africa. This lower eastern section corresponds in the main to the lands of the Somali and Galla tribes, although some Gallas live also on the higher table-lands to the west.

The Abyssinian Highlands.—No other region of equal extent in Africa has such a high average elevation as the Abyssinian highlands, although no one mountain rises to such a height as the snowy peaks of East Africa farther south. One of the most striking features in this high region is the sharply-defined line by which it is bounded on the east. Coming from the shores of the Red Sea, the traveller passes at first over gently rising plains, until he arrives at a wall of mountains stretching in front of him, and mounts by one steep climb to a height of 6000 to 8000 feet. This range extends north and south some 600 miles, coming close to the sea towards the north. It rises here and there into peaks, but more often forms merely the outer edge or cliff of the plateau. In the west the edge is not so well marked, but the plateau still appears from the plains east of the Nile as a range of mountains. The greater part of the highlands bounded by these two outer ramparts is composed of vast sheets of lava and other volcanic rocks poured out from below the surface in past ages. The action of running water has so furrowed the ancient surface of the country that it is now broken up into a sea of ridges and isolated plateaux, between which the streams flow in deep gorges, sometimes hardly accessible to man. In Abyssinia proper there are two principal groups of high mountains 100 to 150 miles away from the eastern rampart. In the north the Simen group rises in Ras Dajan, Abba Yared, and other peaks to 15,000 feet; and farther south the peaks of the Gojam group are almost as high. In the Galla countries south of Abyssinia the land is less rugged and the original form of a plateau can be better seen. The eastern bounding range bends here towards the west, and though at present little known, seems to become higher to the south of a gap formed by the course of the river Hawash. The peak of Wosho, between 6° and 7° N. latitude, is said to be between 16,000 and 17,000 feet high.

River Systems. Towards the western side of the Abyssinian plateau there is a nearly circular area lower

than the general surface, but still over 6000 feet above
the sea. It is occupied by Lake Tsana, or Tana, a deep,
mountain-girt lake, but bordered in parts by low alluvial
land, which must once have been covered by its waters.
Its principal feeder is the Upper Abai River, which rises
almost due south of the lake, and enters its south-west
shore after a course of about 70 miles. The lake is also
fed by copious streams from all sides, so that the Abai,
where it leaves the lake (at its southern point, not very
far from its place of entry), has much increased in size.
It now takes a wide sweep, south-east, south-west, and
north-west, almost encircling the Gojam Mountains, and
flowing, with many a fall and rapid, in a deep, narrow
valley, into which numerous side valleys open. In this
section it takes the name Bahr-el-Azrek or Blue River,
whence it is known generally as the Blue Nile. Finally
it leaves the mountains and flows quietly across the
plains before joining the other branch of the Nile at
Khartum.

The northern parts of Abyssinia are occupied by the
upper branches of the Atbara (the lowest Nile tributary).
The largest of these is the Takazze, which flows round the
north side of the Simen Mountains in a deep and wonderful gorge, whose sides, where not too steep, are clothed with
impenetrable forests. Like the Blue Nile, the Atbara,
after receiving its various branches, flows across the plains
before joining the Nile at the southern edge of the desert.
As we shall see presently, the deep valleys of these streams,
which in time of flood are hardly to be crossed, play an
important part in dividing the country into separate
provinces.

The parts of the plateau south of Abyssinia are
furrowed by streams, which unite in a large river flowing
south in a narrow valley, finally named the Omo. No
traveller has yet succeeded in tracing this river to its
termination, and it is still uncertain whether it continues
southwards, entering the large Lake Rudolf or Basso
Norok, discovered in 1888 by Count Teleki, or makes a
sudden bend either to the east or west. Lake Rudolf
occupies a depression, by which the Ethiopian highlands

are broken to the south, and seems to belong in part to the great East African furrow which will be spoken of in the next chapter.

Climate, Flora, and Fauna.—Owing to their elevation, the Ethiopian highlands enjoy a moderately cool and healthy climate. Their mountain ranges serve to condense the moisture from the winds which blow from the south when the surface of North Africa becomes heated during the summer of the northern hemisphere. Rain falls in torrents, filling the rivers and making them carry down immense quantities of earthy matter, so that they supply to the Nile both the water which inundates its banks and the fertilising sediment which is left when the floods retire. The rest of the year, on the contrary, is nearly rainless. According to the height above the sea, Abyssinia has been divided into three zones of climate and vegetation. (1) The tropical zone, occupying the outer slopes to a height of about 5000 feet, and supporting a vegetation similar to that of tropical Africa generally, forming dense forests in parts; the cultivated plants include cotton, coffee, the sugar-cane, indigo, bananas, etc. (2) The temperate zone, with a climate resembling that of South Europe, embracing the greater part of the country between 5000 and 9000 feet, and producing the cereals and fruit trees of temperate climates, including wheat, barley, and maize, in addition to a native grain called tef; excellent pasturage for cattle is found in this zone. (3) The cool region, above 8000 feet, with few cultivated plants, and a generally scanty vegetation. The two upper zones have a wild vegetation, which is to a large extent common to Abyssinia and the other high mountains of East Africa. The temperate zone has some forests, and junipers and other conifers occur, some growing to a large size. The highest summits have an Alpine flora. Wild animals are seen principally in the lowest zone, and here most of the usual African forms, including the elephant, occur.

The Coast Plains of the Red Sea. Between the Abyssinian highlands and the Red Sea a fairly level plain extends, narrow in the north, where the plateau escarpment approaches the sea, but widening out southwards to almost 200 miles. This is low by comparison with the adjacent highlands, but still reaches a height of 1200 to 1500 feet in parts. The surface is covered principally with scrub, with a sprinkling of larger trees, such as baobabs and sycamores. Water is generally scarce, most of the streams which descend the outer slopes of the highlands becoming absorbed by the soil soon after reaching the plains. The principal stream is the Hawash, which descends from the high plateau, its valley forming the southern limit of the eastern escarpment. On reaching the lower level the Hawash turns abruptly northwards, turning afterwards north-east, and becoming engulfed in a depressed saline area (said to be below sea-level) lying inward from the head of the Gulf of Tajura, which indents the coast for some distance. Another salt plain, also depressed, occurs farther north in about $14\frac{1}{2}°$ N. latitude, and near this are some volcanic peaks, which have been active within recent times.

The Somali and Galla Lands.—At the point where the Hawash valley forms a cleft in the Ethiopian highlands an important ridge of high ground branches off from the heights on the south side of the river, and continues with few breaks in a direction slightly north of east through the whole of the eastern "horn" of Africa, ending at Cape Guardafui, the most easterly point of the continent. This ridge forms the highest ground in the whole area lying south-east of the Ethiopian highlands, and as it keeps close to the northern coast (washed by the Gulf of Aden), the whole country has a general slope to the south-east towards the shores of the Indian Ocean. This high ridge falls abruptly to the north in steep cliffs, and for a great part of its length consists of two parallel chains of mountains with a depression between. In the Harar group of mountains it reaches a height of 6000 feet, and near 45° E. longitude the peak of Gan Libah is said to be considerably higher. The slope of the southern plains is so gradual that their general character is that of a nearly level plateau, broken occasionally by rocky peaks and ridges, and furrowed by the beds of streams.

Climate and Flora.—Except in the west, where the neighbouring Ethiopian highlands condense the moisture from the prevalent south-west winds, this region has but a scanty rainfall, and, except in the neighbourhood of streams, the surface is mostly bare and covered only with stunted mimosa or aloe jungle. Near the western highlands, however, tropical forests are met with, and in the valleys of the northern coast ranges vegetation is more luxuriant. One of the most fertile regions of Somaliland is that known as Ogaden, north of the Webi Shabeli, whose rich pastures support a noted breed of camels, and where durra and other crops are grown. On the whole, the country becomes barer from west to east, and towards the end of the projecting "horn" the region known as Nogal, "the stony land" (in distinction to the "Hand" or "stoneless land" farther west) is almost a waterless desert. The whole of the east coast from Cape Guardafui to the equator is barren and inhospitable, hardly a single stream reaching the sea.

Rivers.—The two principal, in fact the only real rivers of this region, are the Webi Shabeli (Leopard River) and the Jub. Both have their sources in the south-eastern border of the Ethiopian highlands, and derive from them a considerable volume of water, which enables them to cross the arid plains of Somaliland without drying up, though in the dry season they shrink greatly. The Webi Shabeli turns south-west on approaching the sea, and for over 150 miles flows parallel to the coast, until it loses itself in a small lake or swamp only a dozen miles from the sea. The Jub, which has a larger number of important tributaries, enters the sea almost on the equator. Its lower course is navigable by small vessels for a part of the year, but it is very winding and shallow in places, and the mouth is obstructed by a dangerous bar. Rapids occur in about $2\frac{1}{2}°$ N. latitude, at which the current sweeps among the rocks at such a rate that navigation is impossible.

Products. Coffee is the most valuable product of

this region, which is its native home, the name even being

Fig. 21.—Scrub Vegetation in Somaliland.
(Photograph by Captain H. G. C. Swayne.)

possibly derived from the district of Kaffa, south of

Abyssinia. In Somaliland many kinds of aromatic shrubs grow, which have made the country famous since very early times as the land of aromatics. These include myrrh and frankincense, as well as other gums and resins. Wild animals are abundant in these countries. The elephant is still found almost everywhere except in the east, and supplies ivory for exportation. The rhinoceros, giraffe, various antelopes, the ostrich, and crocodile also occur.

2. *Inhabitants of North-East Africa*

Races of Abyssinia.—The Abyssinian highlands are characterised by an extraordinary mixture of races, a fact which is set forth in the very name of the country, for Habesh, the original form of Abyssinia, is said to mean "mixed." Semites, Hamites, and Negroes are all found in the country, though the Negroes are chiefly confined to its outer fringe on the west and a small area in the north. The Semites are represented by the Himyarites, who crossed over from Arabia within historic times, and have since formed, as a rule, the governing race, and a few Arab tribes on the east and west borders. Scattered among them are various Hamitic tribes who once probably formed the bulk of the population, and others who seem to have mixed with the Semites. The Agao Hamites have given their name to the district of Agaomeder, south of Lake Tana, which is even now peopled almost exclusively by them. All these Abyssinian tribes are both cattle-rearers and agriculturalists, the nature of the country being equally suited to both occupations.

Hamitic Races. In the more southern highlands and in the lower plateaux the people are almost entirely Hamites, though with some mixture with the other races on the border lands. They fall into three groups: the Afar or Danákil, between Abyssinia and the Red Sea; the Somali, occupying the projecting "horn" east of a curved line drawn from the Gulf of Tajura to the Lower Jub; and the Gallas, dwelling west of this line. Except quite in the south, all these peoples are nominally Mohammedans.

The Gallas are the most numerous, being supposed to number from 7,000,000 to 8,000,000; the Somalis come next, and the Afar, whose country is much smaller and extremely arid, are the least important. Broadly speaking, the Somalis are distinguished from the Gallas by their darker skin and greater height. They are a wild, lawless race, mostly of nomadic habits, as their country produces little but scanty pasturage for flocks. They are divided into numberless clans, with little tendency to political combination. In the interior the largest villages are those of Logh and Bardera, on the east bank of the Jub; Barri, on the middle Webi Shabeli; and Gelidi, on its lower course. The coast towns will be spoken of below. The Gallas, who generally call themselves Oromo or Oroma, are more cultured than the Somalis, and those of the purest type have exceedingly fine forms and features. They are in the main herdsmen, like the Somalis, but there is among them a distinct class of tillers of the soil, their country being in parts favourable to agriculture. There are several distinct tribes or sections of the Gallas, some of the most important being the Arussi and the Borans, who inhabit the wide plains in the east and south of their domain. In the north some of the tribes have established small kingdoms, as *e.g.* that of Kaffa, in the basin of the Omo. They have also settled in considerable numbers in the south of Abyssinia, especially in Shoa, between the Hawash and the Blue Nile.

3. *Abyssinia*

As has been said already, the deep valleys of its rivers naturally divide Abyssinia into distinct provinces, inhabited by different tribes, and often only loosely united under the Negus or chief ruler. The province of Amhara, stretching from the Takazze valley in the north to that of the Abai, after it has left Lake Tana in the south, has long formed the centre of the kingdom; while Tigré, in the north, and Gojam and Shoa, in the south, have often been governed by half independent chiefs. Within recent years the rulers of Gojam and Shoa have extended their sway over the Galla highlands as far as Kaffa in the south, and over the province of Harar—formerly under a chief of its own, and occupied for a time by Egypt before the Mahdi's rebellion—in the east. On the death of King John in 1889, Menelek, king of Shoa, seized the throne of Abyssinia, becoming supreme over the various provinces. He has lately held his own successfully against the Italians. The religion of the country is nominally Christian, the people having been converted in the fourth century. The form resembles that of the Eastern Church, and therefore Russian sympathy has been aroused, but the worship is very debased, and some pagans are found in outlying parts.

Towns, Trade, etc.—Almost all the chief towns of Abyssinia lie on elevated plains away from the main streams, the deep valleys of which, besides being generally unhealthy, have rather hindered than helped communication between different parts of the country. Many of the towns have at one time or another been capitals of the country, and others are chiefly important as markets and trade centres. Debra Tabor, one of the royal residences, occupies an important position east of Lake Tana, at a height of over 8000 feet above the sea, on the principal line of communication from north to south. Gondar, generally considered the capital, 6500 feet above the sea, lies north of Lake Tana, and is the religious centre of the country. Magdala, a strong fortress near the eastern edge of the highlands, was captured by the British expedition in 1868. Axum, in Tigré, was the capital of the old Himyarite Empire, but is now decayed. Ancient ruins are to be seen in its neighbourhood and in many other parts of Tigré. The present capital of that province is Adua, an important trade centre. Ankober, the capital of Shoa, lies at the great height of 9725 feet above the sea, on the outer slope of the eastern border range. Its position is important as commanding the most direct trade route from the coast to the Galla countries. In Abyssinia proper the places most important as markets are: Korata, on the south-east shore of Lake Tana, and Sokota, near the Upper Takazze, both in Amhara; and Baso, in the south of Gojam,

near the Abai. In the Galla countries important markets
are held at Bonga, the chief town of Kaffa, and near Sobso,
in the small principality of Licka, which occupies a plain
near the head-waters of the Omo. The importance of
Licka consists in its central position between the markets
of Gojam, Shoa, and Kaffa, and its vicinity to some gold-
producing districts to the west. Another place of importance with respect to trade is Harar, a fortified town on
the line of heights which run eastwards through Somaliland. Its position gives it the command of the best route
between the Galla States and the Gulf of Aden, the use of
which is, however, not in favour with the Abyssinians.
As it lies on the head streams of the Webi Shabeli, its
position is also favourable with respect to the Somali
countries to the south, and its neighbourhood produces
abundance of excellent coffee.

At present the most important trade route is that
running north and south from Massaua, on the coast, by
Adua and Debra Tabor, and serving as the outlet for the
commodities of the southern markets also, the most important of which are gold, ivory, and coffee, European
articles being imported in exchange. This route represents
to some extent the ancient trade route of the Axumite
Empire, which led to the coast at Adulis (now Zula), a
little south of Massaua. In spite of the capabilities of
the country, the foreign trade of Abyssinia is but small.
Besides the articles already mentioned, hides, butter,
honey and wax, myrrh and gums are the principal exports.

4. *Italian Possessions*

Eritrea: its History.—In 1870 the Bay of Assab,
at the southern end of the Red Sea, was purchased by
an Italian Shipping Company, and it was declared an
Italian protectorate in 1882. In 1885, when Egypt
abandoned the coast-lands south of Suakin previously
in her possession, the port of Massaua was occupied by
Italy with the approval of Great Britain, and it has since
become the capital of the Italian Red Sea territory, which

includes all the coast from a point about 100 miles south of Suakin to Raheita, just north of the Straits of Bab-el-Mandeb. The name Eritrea, lately given to this territory, is derived from a Greek word meaning "red," applied in ancient times to the Red Sea and Gulf of Aden.

By treaties with local chiefs, Italy soon afterwards gained a footing on the coast south of Cape Guardafui, and the greater part of the Somali and Galla countries as far as the Jub River, together with Abyssinia, were in 1890 recognised by Great Britain as an Italian protectorate, some positions on the coast, which had belonged to the Sultan of Zanzibar, being ceded by him in 1892. But although King Menelek concluded a treaty of friendship with Italy in 1889, he has since successfully resisted all attempts to make Abyssinia a vassal state. The northern part of Tigré has, however, been brought under Italian influence, and the important town of Kassala, on the western plains near the Atbara, has been temporarily occupied with the consent of Egypt.

Massaua. Trade, etc.—Massaua owes its importance to the fact that it is the natural outlet for the trade of Abyssinia, which has, however, been hampered by the fact that for centuries the port has been in the hands of outsiders unfriendly to Abyssinia. The town, which is fortified, lies on an island, and its port—one of the best on the Red Sea—occupies the sheltered space between it and the mainland. Its trade, which has been referred to under Abyssinia, is mainly in the hands of Indian and Greek merchants. The population is a little under 8000, about 600 being Europeans. The neighbouring sandy plains of the mainland make the place exceedingly hot, but among the hills at a little distance from the coast the climate is pleasanter. Agriculture in the colony is hindered by scarcity of water, and the native population is chiefly pastoral and nomadic. On the uplands, however, wheat and barley have been grown with success.

Somali Coast. The once important Arab marts of Magdoshu and Brava on the eastern Somali coast have now sunk into insignificance. The Italians, however, have some settlements on the coast, and a trading company has been formed which has met with some success. The administrative centre is at El Adhale (now called Itala) in about 3° N. latitude.

5. *British Possessions*

Somaliland.—The occupation of a part of the Somali coast by Great Britain was part of a general policy of securing the overland route to India from foreign menace, and the protectorate is in a measure subordinate to Aden, itself administered by the Government of Bombay. As far back as 1840 and 1858 the Indian Government secured the rights over some small islands at the west end of the Gulf of Aden, but the whole coast was afterwards occupied by the Egyptians during the extension of their dominions in the years following 1870. In 1884 Egypt was forced to abandon her posts on this coast, and they were formed into a British protectorate, the limits of which have since been defined by agreements with Italy and France. On the coast it stretches from Ras Jibuti, on the south of Tajura Bay, the greater part of the distance to Cape Guardafui, and for a good part of this length it has a breadth of from 170 to 220 miles, reaching to 8° N. latitude. The principal outlets of Somaliland—Zaila, Bulhar, and Berbera—are all in this territory, and since the British occupation trade has much increased, and would do so still more if the ports could be brought into relations with the Galla countries. Berbera is the largest of the three towns, containing 30,000 inhabitants when the caravans of camels arrive from the south and the natives from the neighbouring districts flock in to trade. The principal exports are cattle and sheep, hides, ostrich feathers, and gum, while rice, cotton goods, and dates are imported. A large amount of coffee from Harar has recently been exported *via* Zaila.

The Island of Sokotra, lying off the eastern extremity of Somaliland, was occupied by the Portuguese during the palmy days of their Indian Empire, and troops were also stationed there by the East India Company in 1835, but it did not definitely become a

British possession until 1886. Politically it is a dependency of Aden. The only place of importance is Tamarida, on the north coast, the port of which is frequented by Arab vessels (dhows) from the coast of Arabia. The principal products of the island are cattle, and the kind of aloes named from it Socotrine.

6. *French Somali Coast*

The French hold the country bordering on Tajura Bay, where they first established themselves in 1855 at Obok, on its northern shore. This has lately been replaced as the centre of administration by the port of Jibuti, on the southern shore of the bay. The colony has few resources, and the hopes entertained of attracting the trade of Harar and Abyssinia to the place do not seem likely to be realised. It is chiefly important as a strategic post on the route to the East.

CHAPTER X

EAST AFRICA

1. *Physical Features*

General Characters.—With the exception of a strip of lowland bordering on the coast, the whole of East Africa, with which we are concerned in this chapter, consists of a broad high plateau with an average elevation of about 4000 feet above the sea, and attaining a breadth of 700 to 800 miles from east to west. In the north it is to some extent continuous with the Abyssinian highlands, from which, however, the depression of Lake Rudolf supplies a natural line of separation. These East African highlands are characterised, above all, by the number and large size of the lakes, which occupy the lowest parts of depressions running, as a rule, north and south. This north and south direction is to be seen, too, in the general trend of the mountain ranges; so the whole region may be said to be made up of parallel lines of mountains, table-lands, and hollows ranged side by side. On the east the ground rises in two or three distinct steps from the coast lowlands, and in the west the plateau falls somewhat rapidly, but without any decided step, to the great forest-clad basin of the Congo in West Africa. Southwards it is divided from the equally high plateau of South Africa only by the valley cut by the Zambesi in its course to the sea.

In the northern parts of East Africa the principal lines both of heights and hollows lie not far from the eastern and western edges of the plateau, while the central space is occupied by a broad upland with a very uniform surface. This consists chiefly of ancient crystalline rocks (granite, etc.), while the bordering ranges, especially on the east, are remarkable for the number of volcanic peaks which have been thrust up in later geological times, and which are found, as a rule, not far from the great troughs or valleys which furrow the surface. The formation of both seems to be due to forces acting from below the earth's crust.

The Eastern Heights and Hollows.—From the lowlands on the coast the plateau rises in a series of steps. Near its eastern edge there is to be seen in places an outer line of hills formed of gneiss and other crystalline rocks, of no great height, but representing the ancient line of highlands which traversed East Africa from north to south. Between these heights and the damp coast plain a dry, barren plain of moderate height occurs, known sometimes as the Nyika. Westward of the first ranges, the plateau increases in height, and forms a broad zone, much of it over 5000 feet above the sea, and widening out northwards to a breadth of 200 miles. A large part of the surface is composed of volcanic rocks, and at intervals rise the great volcanic peaks, mostly extinct, which are the highest mountains of all Africa. The two best known are Kilima Njaro, in the south, and Kenya, farther north, almost on the equator, both rising to 19,000 feet or more, near the eastern edge of the high plateau; but many other peaks and ridges are undoubtedly volcanic. These highlands are divided into two by the great East African trough or valley, which runs southwards from the direction of Abyssinia, and extends without a break to about 6° S. latitude. It is thought to have originated in a great crack or subsidence in the earth's crust in former ages. Its sides are formed sometimes of steep cliffs, rising in one or more steps to the plateau, and at others by broken ranges of hills. The floor is in parts level, but sometimes broken and hilly, and it is divided into separate sections by transverse ridges. The altitude of the floor varies greatly, rising to over 6000 feet at about 1° S., where the height of the plateau also is greatest. A number of lakes, large and small, all of them without outlet, lie in this valley, the principal being Lake Rudolf in the north. Towards the south a branch valley runs to the south-west, also containing a lake (Eyasi).

The Western Trough.—Near the western edge of the plateau another zone of high ground occurs, forming the east and west sides of another great longitudinal valley, the lakes of which are larger even than those of

FIG. 22.—ELEPHANTS IN THE "NYIKA" STEPPE.

the eastern valley, quite half of the length being covered by water. As it occurs not far from the central line of the continent, it has been called the Central African trough. In the north it starts from the valley of the Upper Nile, and passes southwards by the two western Nile lakes, continuing in this direction until it reaches a point far southward of the southern end of the eastern valley. The sides are, as a rule, steep and high, and, both east and west, often rise to a height of over 5000 feet above the sea. The principal mountain ranges occur on the east. On this side, between the two Nile lakes, just north of the equator, is the great snowy range of Ruwenzori, volcanic in its origin, like the more eastern peaks, but made up of a larger number of ridges and summits. It forms an oval-shaped mass. The most lofty summit is probably nearly 18,000 feet in height. Farther south the great trough is bordered on the east by an almost continuous line of highlands, rising in parts into regular mountain ranges 8000 to 10,000 feet high. South of the Nile lakes a line of volcanic peaks crosses the floor of the trough, which here reaches its greatest height above the sea, forming the water-parting between the streams flowing north and south. These peaks occur almost in the same latitude as the highest part of the eastern trough, which has likewise volcanic cones rising from its floor. They begin on the east with the peak of Mfumbiro, first seen from a distance by Speke in 1861, and end in Kirunga, still partially active and remarkable (as a volcano) for its distance from the sea (700 miles).

The Southern Section.—In the south these two lines of heights and hollows are not found. The eastern line becomes less and less marked, while the western, though partially interrupted for a space, seems to bend eastwards and afterwards to occupy a more central line. This easterly bend of the trough occurs between the two great lakes which occupy so large a part of its floor, and it is here represented only by clefts cut by streams in the highlands. But the absence of a well-marked depression on the direct line is made up for by a short parallel hollow a little distance to the north-east, which is apparently a similar line of subsidence, and is occupied by the narrow lake, Rukwa, similar to those in the main trough. After resuming once more its southerly direction the trough is again well defined, and is still bordered by highlands on each side; but here again the eastern line, which rises in the Livingstone Mountains to 11,000 feet, is the most important. From this range a minor line of heights seems to branch off in the direction of the main eastern line farther north.

In its main features the East African plateau has therefore somewhat the form of a letter Y, the space between the two branches having a generally uniform surface. It is, however, not entirely isolated, for a broad zone of high ground (3000 to 5000 feet) branches off towards the south-west, extending across the continent to the west coast, and uniting the East African with the South African table-land.

Lakes and Rivers.—In East Africa the rivers are quite eclipsed in importance by the great lakes. In no other part of the world except North America do so many large lakes occur. They are of two distinct types, the narrow with steep shores, and the broader with gently sloping margins. The narrow lakes all occur in parts of the great valleys above described, especially in the western of these. The centre of the western valley is occupied by Lake Tanganyika, the longest fresh-water lake in the world (almost 400 miles), and the southern end by Lake Nyasa (nearly 350 miles). The breadth varies from 15 to 50 miles, so that they might almost be called inland seas rather than lakes. Both are very deep. Tanganyika, which is 2750 feet above the sea, was long thought to have no outlet, but it was at last found that (periodically, at least) its surplus water escapes through a gap in the western wall towards the Congo. Its chief feeder is the Malagarazi, which, with its many branches, drains the central part of the plateau and reaches the lake through a gap in the eastern line of heights. North of Tanganyika, and draining into it by the Rusizi River, is the smaller lake, Kivu (3200 feet). Lake Nyasa (1300 feet) sends out a river, the Shiré, at its south end, which joins the Zambesi. Its principal feeders come from the west, but none are of large size. To the east of the south end is the smaller lake, Shirwa, which has no outlet, and seems to be shrinking. The only lake of the eastern valley which can at all compare in size with those just mentioned is Lake Rudolf (Basso Norok of the natives), at its northern end. Only the narrower southern half has the typical high shores of the valley lakes, for towards the north, where it widens considerably, the country is much lower on each side. The water is slightly brackish, and the lake seems to have no outlet. The only important feeder is the Nianam, at its northern end, thought by some to be the Omo of the Galla countries. East of its northern end lies a smaller lake (Basso Naebor or Stefanie), which has no outlet, and appears to be brackish or fresh, according to the height of its waters. The other lakes in the eastern valley are mostly small, and are all without outlets to the sea. The most important from north to south are Baringo, Naivasha, and Manyara in the main valley, and Lake Eyasi, situated in a south-western branch of the valley, and receiving the drainage of a part of the plateau to the west. It varies greatly in size, according to the season. The principal representative of the broad lakes is the Victoria Nyanza, the largest in Africa, which lies in a hollow of the central part of the plateau. It has already been spoken of in describing the Nile sources. Two other lakes occur on the south-western extension of the plateau. Bangweolo is a round shallow lake, about 60 miles in diameter, with swampy reed-covered shores, and Mwero, farther north, seems to hold an intermediate position between the two types. These both belong to the Congo system.

Except on the outer slope of the plateau, almost all the rivers flow into one or other of the lakes above mentioned, and thus belong either to the systems of the Nile, Congo, and Zambesi on the one hand, or on the other to an extensive area of inland drainage in the east, containing those lakes which have no outlet. The larger lakes are admirably adapted for purposes of navigation. Those of the narrow type, in particular, owing to the great distances over which they extend, supply an excellent natural waterway through the heart of the continent. The rivers of the plateau, on the contrary, are of little use for navigation, and almost the only one deserving mention in this respect is the Kagera, which flows into the Victoria Nyanza from the south-west.

The Coast-lands and their Rivers. The country watered by the coast streams includes, in addition to a narrow strip of lowland close to the sea, the outer steps of

the central plateau up to a height of 2000 to 3000 feet.
The width of this zone varies greatly in different parts.
Opposite the central point of the coast the high plateau
comes nearest to the sea, but it retreats from it both
north and south, leaving a much wider space for the development of rivers. This is due also to the outward
sweep which the coast itself makes in each direction. The
principal river in the north is the Tana, which derives a
considerable amount of water from Mount Kenya and the
neighbouring ranges, from which its head streams flow,
but which receives not a single tributary worth mentioning
in its passage across the lower grounds. It is a winding
river with a strong current, and abounds in snags (or
trunks of uprooted trees), which make navigation difficult,
though possible for about 200 miles from the mouth.
The Sabaki, known in its upper course as the Athi, is
almost as long as the Tana, but is rapid and useless for
navigation. In the south there are two fairly large rivers,
the Rufiji and Rovuma, each formed of several branches,
so that the widest parts of their basins are near their
sources. The Rufiji derives its water from the mountains
near the head of Lake Nyasa and other ranges running
thence to the north-east, which condense the moisture from
the south-east trade wind. The Ulanga, one of the chief
head streams, flows in a level swampy valley, which holds
the water, so that there is a good supply at most times
of the year. The course of the Rufiji is broken at two
places by falls, but above and below these it is navigable
for small craft, as it flows chiefly through level plains.
It forms a considerable delta at its mouth. The head
streams of the Rovuma take their rise in the line of
heights which bound Lake Nyasa on the east. The
southern branch, the Lujenda, rises in a swamp, separated
from Lake Shirwa only by a ridge 50 feet high, over
which the lake water must once have found an outlet to
the river. The Lujenda is composed of still reaches,
separated by rapids. In its central course the Rovuma
flows through a wide plain dotted over with isolated hills,
and spreads out to a considerable width, but is shallow
and much blocked by sand-banks, so that it is of little use

for navigation. Before it reaches the sea the plateaux on either side close in, and it flows through a narrow valley bounded by steep sides.

Several islands lie at a little distance from the coast, the most important being Zanzibar (a little over 50 miles long), placed just at the centre of the inward curve of the coast, and Pemba, farther north. Zanzibar is traversed by a range of sandstone hills, but much of the surface is very fertile, as is also that of Pemba. Both are parallel to the coast and about 30 miles distant.

Climate, Flora, and Fauna.—The climate, and with it the vegetation, of East Africa varies according to the height above the sea. The coast lowlands have naturally the highest temperature (average over 80° Fahr.), while on the higher parts of the plateau the mean temperature for the year falls below 70°. Rain occurs most plentifully near the sea and in the neighbourhood of mountains exposed to the sea breeze farther inland. Apart from altitude, however, there is a considerable variation in the rainfall, for, speaking generally, it becomes scanty in the north-east, where the comparatively dry region of the Somali and Galla countries is approached, while the western borders touch on the rainy region of West Africa. On the hot and moist lowlands on the coast a rank tropical vegetation similar to that found generally on the shores of the Indian Ocean prevails. The lower steps of the plateau, which include the dryest part of the whole region, are mostly covered with a scrub vegetation mixed with stunted trees, euphorbias, or thorny jungles. During the dry season the trees are often bare of leaves, and present an almost lifeless aspect. One of the most inhospitable districts is that stretching south-west of Kilima Njaro as far as Ugogo, where a plain, burnt up and waterless in the dry season, but a miry waste during the rains, is with difficulty crossed by caravans. In the more fertile districts the alternation of grassy expanses with clumps of trees often brings about a park-like appearance. The higher plateaux are mostly covered with rich pastures, serving admirably for cattle rearing, and sometimes with forests. On the mountain slopes and isolated plateaux exposed to the influences

of the moist sea breezes a dense and luxuriant forest is
found, which is supposed to have once covered a much
larger area than at present. The higher mountains are
marked by a regular succession of zones of vegetation,
which are very similar even in the case of mountains far
removed from each other, whose upper flora, resembling

Fig. 23.—Park-like Scenery in East Africa. (After Stuhlmann.)

that of the Abyssinian mountains, is nowhere represented in the intervening country at the present day.
The lower slopes are generally bare of trees, except along
the courses of streams. Higher up a continuous belt of
forest stretches round the mountains, merging upwards
into a zone of bamboos. Higher still the slopes become
grassy and open, and an Alpine flora appears, which
includes some strange forms, such as tree lobelias and
Senecio (a gigantic kind of groundsel), while on the
highest mountains of all, perpetual snow covers the
summits.

The banana is largely grown in parts of East Africa,
forming sometimes, as in the country west of the Victoria

Nyanza, continuous groves over large areas. Of cereals,

Fig. 24. Bamboo Forest, Mount Elgon.
(Photograph by Mr. Ernest Gedye.)

millet is the most important, and other cultivated plants

are cassava and the sugar-cane. Most of the usual animals of tropical Africa are met with, and in the open grassy steppes, zebras, antelopes, giraffes, etc., occur in large herds. The elephant, which not long ago was plentiful over the whole region, has been so persecuted for the sake of its tusks that it has been exterminated over a great part of the area, particularly between the coast and Lake Tanganyika.

2. *Inhabitants of East Africa*

As East Africa includes the borderlands between the Hamitic and Bantu races, its inhabitants are as varied as those of any other division of the continent. As elsewhere, the Hamites seem to have been pushing southwards for many centuries, and though, with the exception of some Southern representatives of the Gallas and Somalis, no undoubtedly pure Hamite stock is found, Hamite influence has evidently been at work in a large part of the country towards the north. The Masai, who inhabit the broad uplands between the coast and the Victoria Nyanza, are the most decidedly Hamitic in type, being tall and well built, but they speak a tongue akin to that of the Negroes on the Upper Nile. In the steppe region, a little farther south, Bantu tribes and others apparently Hamitic are scattered over the country without any definite dividing line, whilst the remnants of an aboriginal race, distinct from either, have also been traced by some observers. In the west and south of East Africa the population is almost entirely Bantu, but scattered among them representatives of a lighter race of finer physique, known as the Wahima or Watussi, are found. Their language shows them to be allied to the Gallas. In some parts they actually form the ruling race, in the same manner as the Fulahs in West Africa lord it over the Negroes. Even among the Bantu, though the various dialects spoken are so generally alike that one tribe has no difficulty in understanding the speech of another, there is much variety in physical appearance. Towards the south the country has been subject to raids from tribes allied to the Zulus of South Africa, who, though Bantu, play much the same part here as the Masai do farther north. They are found chiefly in the districts west of Lake Nyasa and near the southern end of Tanganyika. On the coast a population of Arabs and Hindus has settled, and the great numbers of slaves from all parts of East Africa (with whom the Arabs have intermarried) add to the confusion of races. The mixed native population is generally known by the name Swahili. The Arabs have pushed as traders in ivory and slaves over the whole of the interior, and have brought about a profession of Mohammedanism in many parts. The number of European immigrants is quite inconsiderable.

Modes of Life.—The slight tendency on the part of the Negroes to form powerful states is nowhere better seen than in East Africa, where other causes too have acted in the same direction. The inroads of the Masai and other Hamites from the north, and of the Zulu tribes from the south, have helped to create a state of unrest and insecurity, but more important still has been the effect of the slave trade, which has existed for centuries in this region. For the purpose of obtaining slaves to sell to the coast-traders, one tribe makes war on another, and the population often becomes split up into a number of hostile sections, the separate towns or villages of a district being often at daggers drawn with each other. Large areas are often depopulated completely by the slave-hunting expeditions. The only State which has, of late years at least, risen to any importance in East Africa is that of Uganda, where the rulers have been the Wahima intruders of Galla origin.

Both cattle-rearing and agriculture are practised in East Africa, even by the Bantu, but, as elsewhere, it is the representatives of the Northern races with whom cattle rearing is carried to the greatest extent. The Masai in particular have been accustomed to live entirely on the produce of their herds, for which the rich grassy uplands of their country were especially favourable. The whole organisation of the tribe is based on the possession of cattle, the male population being divided into elders, herdsmen, and warriors whose business it is to protect the herds and add to their numbers by raids on weaker tribes. But of late years a terrible cattle plague has raged through East Africa, sweeping off the whole wealth of many tribes, and the Masai have been reduced to a state of famine. In fact, with the introduction of civilised government, which would put a stop to their cattle-raids, it has been thought that the Masai are bound to die out in time. In the countries lying round the Nile sources, cattle-rearing is the special business of the Wahima or Watussi, who form a separate class among the Bantu population. They possess a remarkable breed of cattle with enormously large horns, such as is met with also in Abyssinia, from which direction it has probably been brought along with the people themselves. Farther south the Zulu invaders are the chief rearers of cattle.

Of the various Bantu tribes of East Africa, the Wanyamwezi, who occupy a central position between the coast and Lake Tanganyika, are the best known and most enterprising. Following the example of the Arab traders, by whom they have been largely employed, they often make distant trading expeditions on their own account, and, like the Arabs, have been guilty of devastating whole regions visited by them.

3. *British East Africa*

History.—Although by far the largest part of East Africa first became known through the labours of British explorers, who from 1858 onwards made it a special field for their activity, it was not until 1886 that any definite acquisition of territory was made, and this only after a large part of the country had already been appropriated by Germany. The whole coast, with an indefinite extension inland, had been previously regarded as subject to the Sultan of Zanzibar (see p. 51), whose authority, however, in fact hardly extended beyond the coast towns. As soon as a section of the country had been secured for Great Britain by an agreement with Germany, Sir W. Mackinnon (who for some years had been active in his attempts to open this part of Africa to legitimate trade) joined with other merchants and philanthropists to form a company, which obtained an Imperial charter in 1888. Under the terms of this charter the whole administration of the British sphere was placed in the hands of the Company, which at once commenced operations. The area reserved to Great Britain lay south of the Tana River, and extended southwards almost to Kilima Njaro. An arrangement was made with the Sultan of Zanzibar by which the coast-lands, as well as the interior, were placed under the Company's rule. Mombasa, a little north of Zanzibar, the best port on the East African coast, became the centre of administration, and various expeditions were sent to found stations and spread British influence in the interior.

One of these made its way to Uganda, north of the Victoria Nyanza. Here British missionaries had been actively at work since 1877, and in spite of many difficulties arising from the capricious characters of King Mtesa and his son and successor Mwanga, had met with unusual success in obtaining converts. Mohammedan opposition had been aroused and further complications had arisen owing to the arrival of French missionaries, who made numbers of Roman Catholic converts. The jealousies between these religious bodies often resulted in bloodshed. By an agreement with Germany in 1890, Uganda was recognised as within the British sphere, which on the coast also was extended northwards as far as the Jub River. Zanzibar at the same time became a British protectorate. The work of restoring order in Uganda fell to Captain Lugard, who finally overcame the many obstacles in his way, but the Company, on whose resources these distant expeditions were a great drain, found themselves unable to maintain a permanent occupation. After an agent had been sent by the British Government to report on the situation, the country was at last permanently placed under a British Commissioner directly responsible to the Imperial Government. Both under the Company and the Commissioner, British influence has been extended north-west and west to the western Nile lakes. After long negotiations the rights of the

Company over the coast-strip which had been acquired from the Sultan of Zanzibar were purchased by Government, and the administration of the whole of British East Africa thus fell into the hands of Government officials. Zanzibar and the other islands, with the district of Witu, north of the Tana, are still nominally governed by the Sultan under British protection.

Resources, Trade, etc. Zanzibar.—In considering the resources of British Africa it is necessary to make a distinction between Zanzibar and the mainland territory. As the long-established commercial centre of East Africa, Zanzibar possesses an importance in respect to trade which no other port can yet hope to rival. Almost the whole trade between East Africa and foreign countries has been wont to pass through its port, from which the various commodities have been distributed to their final destinations. Some of this trade may in the future proceed at once to mainland ports, but its advantageous position must always make Zanzibar a very important centre. The island is besides, with the others near it, a great centre of the cultivation of tropical products. The city of Zanzibar lies on the west side of the island, facing the mainland. In it the caravans for the far interior have generally been organised before making their start from one of the opposite towns, and it has thus been the virtual starting-point of most of the great exploring expeditions which have penetrated the continent from the east. Its population, composed of Arabs, Hindus, Persians, and slaves from all parts of the mainland, amounts to at least 100,000. The port, which was declared free in 1892, is merely a roadstead sheltered by the island, and large vessels cannot come close to the shore. The principal product is cloves, which are produced in great abundance. The coco-nut palm is also largely grown and *copra* (the dried kernels) is exported. Many other tropical products could, no doubt, be grown on the fertile soil of the island. The greater part of the trade is in the hands of the Banyans (natives of India). India supplies the largest proportion of the imports, while the greatest share of the exports goes to Great Britain, India following next.

The Mainland.—Although possessing great capabilities, the resources of the mainland territory are still undeveloped. Much has been done in the way of introducing settled government, and the country of the once dreaded Masai can now be crossed in comparative safety. The trade has slowly but steadily increased since the British occupation, but the great need of the country is that of improved means of communication, for, with the exception of a short railway near Mombasa, the only means of transport is at present supplied by native porters. A survey for a railway to the shores of the Victoria Nyanza was carried out in 1892 under Government auspices, and the construction has now been begun, the first rail having been laid in May 1896. The proposed route leads mainly north-west from Mombasa, ascending to the high Masai plateau by the upper waters of the Athi or Sabaki River, descending into the great valley near Lake Naivasha, ascending again on the other side farther north, and finally bending south-west to the Victoria Nyanza by the valley of the Nzoia. With proper communication with the coast the cool uplands of the Masai country, on which the cereals of temperate climates could be grown, would, it is thought, be suitable for regular colonies of Europeans, and the lower but fairly healthy region farther west (Uganda, etc.) would be favourable for coffee-planting. Hitherto ivory has been the only important product of the interior, but with the decrease in the number of elephants its value has greatly declined, and must become still less in the future. The coast-lands are unsuited for European residence, but it may be expected that tropical products will be grown by native labour, as in other tropical colonies.

Coast Towns.—Mombasa, the administrative centre of British East Africa, is situated, like so many of the East African ports, on a small coral island close to the mainland. It was in old times an important Arab port, and was occupied by the Portuguese soon after the discovery of the sea route to India. They were expelled in 1698, and until recent years it completely lost its importance. Harbour works have lately been completed, lines of steamers call at its port, and trade is now

Fig. 25.—Mombasa. (Photograph by Sir John Kirk.)

increasing rapidly. Freretown, named after Sir Bartle Frere, on the mainland opposite, was founded as a home for freed slaves. Malindi, farther north, was, like Mombasa, important in early Arab times, but has now but a small trade. Kismayu, just south of the mouth of the Jub, is the chief port of southern Somaliland, in the north of the British territory.

4. *German East Africa*

History.—The first acquisitions of territory in East Africa by Germans were made at the end of 1884, when Dr. Peters, Count Joachim Pfeil, and Dr. Jühlke, some of the most active members of the German colonial party, proceeded to Zanzibar and concluded treaties with various native chiefs on the mainland, especially in the district of Usagara. Protests were made by the Sultan, who, however, at last yielded before a display of force in the form of a naval squadron. In spite of British claims in the region of Mount Kilima Njaro, the German boundary was eventually drawn so as to include the mountain, making a bend northwards for this purpose, and passing thence to the shore of the Victoria Nyanza in 1° S. latitude. In the south an agreement was come to with Portugal by which the course of the Rovuma, and a line from the mouth of its tributary, the Msinje, to Lake Nyasa, were laid down as the southern boundary of the German possessions;[1] and as the Sultan soon afterwards leased all his possessions on the coast south of the British sphere to Germany, that country became supreme in the whole region between the British and Portuguese possessions, having a length from north to south of 750 miles. Towards the interior the limits were at first left indefinite, but in 1890 they were extended to Lake Tanganyika in the west; a line from Tanganyika to Nyasa in the south-west; and 1° S. latitude in the north-west. A revolt of the coast tribes, headed by the Arabs, gave the Germans much

---

[1] A small portion of the coast south of the Rovuma, between it and Tungi Bay, formerly a part of the Sultan's possessions, was afterwards assigned to Germany.

trouble in 1888-90, but it was at last put down. Although a station (Bukoba) has been founded as far inland as the west coast of the Victoria Nyanza, the German occupation is at present effective only in the districts nearer the coast.

Towns and Trade Routes.—There are several towns on the coast of German East Africa which have long been centres of Arab trade and intercourse with the native populations. One of the great trade routes of East Africa traverses the German sphere from east to west, starting from Bagamoyo, opposite Zanzibar, and passing by the Arab mart of Tabora, in Unyanyembe (whence another route branches off to the north-west), to Ujiji, on Lake Tanganyika. This has been the line by which slaves and ivory—the only products of the country in the past—have been brought down to the coast from the remote districts in the interior. With the recent decrease in the amount of ivory its importance has declined. Bagamoyo has no good harbour, and has owed its importance to its position opposite Zanzibar. Its rival is Dar-es-Salaam, a little farther south, which has a good natural harbour, and is destined to be the chief port of the German territory. Farther north the chief places are Pangani (at the mouth of the river of the same name, which comes from the southern slopes of Kilima Njaro) and Tanga, at which the German steamers call on their arrival from Europe.

Both these places are starting-points for caravans for the district of Usambara and that around Mount Kilima Njaro. Towards the south of the German coast there are two towns named Kilwa, one on the mainland (Kilwa Kivinja), the other on a small islet off the coast (Kilwa Kiswani). That on the islet is the older, having been an important Persian and Arab centre long before the arrival of the Portuguese, and many ancient ruins are still to be seen.

Resources and Prospects.—The future resources of the country will probably consist chiefly in its cultivated products, together with some natural vegetable products, such as indiarubber, gum-copal, orchilla weed, etc. Cattle may also be reared with success on the grassy uplands. Less of the surface than in British East Africa is high enough to be suitable for permanent colonies of white men, and (with the exception of Mount Kilima Njaro) the most favourable districts lie at a long distance from the coast. Such are the Karagwe highlands, west of the Victoria Nyanza, and parts of the line of high ground east of Lake Tanganyika and north of Lake Nyasa. Of the ground below 5000 feet above the sea the value differs greatly, a large portion towards the eastern edge of the plateau (Ugogo, etc.) being arid and unpromising. Between this and the coast, however, there are several fertile hilly districts, especially Usambara, in the north, and Usagara, just south of the main trade route. The Usagara highlands were the first district to attract the attention of the German colonists by their agricultural capabilities, two stations having been founded there in 1885. The cultivation of tropical products, such as tobacco, cotton, and coffee, has also been carried on with some success nearer the coast. The slopes of Kilima Njaro, above the lower arid region and below the forest belt, enjoy a pleasant climate, and are valuable as a sanatorium, besides offering prospects of successful cultivation to settlers. The district of Konde, at the head of Lake Nyasa, is most fertile, and is near high and healthy uplands. The station of Langenburg has been founded on the lake (on which the Germans have a steamer), and others in the country to the north. At present the chief exports are ivory and the vegetable products above mentioned. Railways have been much talked of as an aid to the development of the country. Two main lines have been proposed, one starting from Tanga, and passing by Kilima Njaro to the Victoria Nyanza; and the other—for which a survey is now (1896) being carried out—following in the main the line of the trade route to Lake Tanganyika, with a branch to the northern lake.

Missions, both German and English (the latter started before the German occupation), are scattered over a large part of the country, and are doing much to introduce civilising influences among the natives.

CHAPTER XI

THE ZAMBESI REGION

THE Zambesi, which on the one hand serves as a natural line of partition between the East African and South African plateaux, may, from another point of view, be regarded as the central channel of a region embracing both sides of its course, to which it affords a natural line of entry from the coast. This region will roughly correspond to the basin of the river and its tributaries, with the adjoining portion of the coast-lands watered by smaller independent streams. With regard to the political relations of the country, the dividing influence of the river has hardly been felt at all, and we may therefore in this chapter consider together those parts of the two high areas which are brought into certain relations to each other by their nearness to the river.

1. *Physical Features*

The Zambesi Basin: Relief.—By far the larger part of the Zambesi basin lies to the north of the east-to-west furrow traversed by the main stream. Not only all the principal tributaries but the head-waters of the Zambesi itself flow down from the band of high ground which branches off from the East African plateau and extends westwards almost to the shores of the Atlantic Ocean. The water-parting between these southward-flowing streams and those flowing northwards to the Congo is rarely formed by well-marked mountain ranges, and towards the west it is sometimes almost indistinguishable to the eye. In the more central parts a broad extent of wooded upland intervenes between the two basins, sometimes rising into ranges of hills. The water-parting takes a considerable sweep southwards, so that the Congo basin reaches a much more southerly latitude here than elsewhere. Farther east the uniformity of the plateau is broken by two great valleys much below the general level—those of the Loangwa, and of the Shiré and Lake Nyasa.

On the south of the main stream the eastern and western sections of the basin show much contrast. In the west the country is generally flat, so much so that some of the streams appear to be connected with both the Zambesi and the inland basin of Lake Ngami to the south. The eastern section, on the contrary (in the Matabili and Mashona countries), is a part of the high South African plateau, and the limits of the basin are marked by a line of hilly ground which runs from south-west to north-east. The Matoppo Hills, in the south-west, form the most continuous range, and a branch from this hilly country is continued northwards as the Umvukwe range almost to the Zambesi. Elsewhere the surface of the plateau is often broken by isolated hills, generally of granite, called kopjes. The eastern edge of the plateau is also marked by broken ground and steep escarpments, from which some of the coast streams take their rise. This hilly region is continued again on the north side of the Zambesi, particularly on the east of Lake Nyasa and its outflowing stream, so that in spite of the fact that the general course of the Zambesi is from west to east, it still holds good here, as in the southern limb of Africa generally, that the highest land lies on the east side of the continent.

The Upper Zambesi.—Of the five principal streams which descend the slopes of the northern plateau, the most westerly but one is regarded as the upper course of the main river. The head-stream (known as the Liba) rises in a hilly district almost in the centre of the width of the continent, and begins its course by flowing almost due west, or exactly the reverse of its subsequent general direction. Near the point where it turns southward there is a level sandy plain (in the district of Lovalé), covered with two or three feet of water in the rains, but quite

Fig. 26.—Scenery in Eastern Mashonaland. Photograph by Mr. W. A. Eckersley.
(From the *Geographical Journal*.)

dry at the opposite season of the year, which is drained towards both the Zambesi and the Congo. After joining another branch coming from the north-east, the river is much increased in size and assumes the name Liambai. During its passage through the Barotse country its valley is bounded by low hills, and the intervening strip is much subject to inundations. Before turning decidedly east a series of rapids occurs, the principal of which, the Gonyé Falls, entirely obstruct navigation. Hereabouts the banks of the river are either rocky or wooded, while wooded islands rise from the broad surface. The beauty of the country much impressed Dr. Livingstone on his first reaching the river from the arid regions farther south. In about 18° S. latitude the Zambesi receives from the west the Chobé, or Kwando (its most westerly tributary), which has flowed during the last part of its course through a level tract subject to inundations, splitting up into several arms. Its head-waters lie in a hilly, well-wooded country intersected by streams, and in its middle course it flows through grassy plains. Shortly after the junction the Victoria Falls are reached—perhaps the most remarkable waterfall that exists. At some not very remote period a great crack or fissure of the surface has been formed, running right across the bed of the river, here over a mile wide. The whole stream plunges into the abyss, which is nearly 400 feet deep, and in places less than 100 yards wide. The only outlet for this vast body of water is a second narrow fissure at right angles to the first, along which the river rushes in a zigzag course in boiling eddies and whirlpools, only escaping into a broader bed many miles below. The vast columns of spray which rise from the abyss, together with the ceaseless thunder of the water, which is heard to a great distance, have given rise to the native name Mosi-oa-tunya, or the Sounding Smoke.

The Central and Lower Zambesi. -The Victoria Falls may be considered to mark the end of the Upper Zambesi. Throughout the whole of its central course the river, flowing north-east and east, is generally rapid and only navigable during high water. Its banks are bounded as a rule by a plain of varying width lying between the plateau escarpments on the north and south. Occasionally these approach so near as to form a gorge. Two large northern tributaries join the river in this central section. The first is the Kafukwe, or Kafue, still little known, which, rising in the wooded hills to the north of the basin, makes a wide sweep westwards before bending east to join the main river. The next is the Loangwa, also flowing westwards in its central course, whose valley forms a wide break in the northern plateau, possibly once occupied by a lake. The floor of the valley is generally flat and is bounded by steep escarpments, such

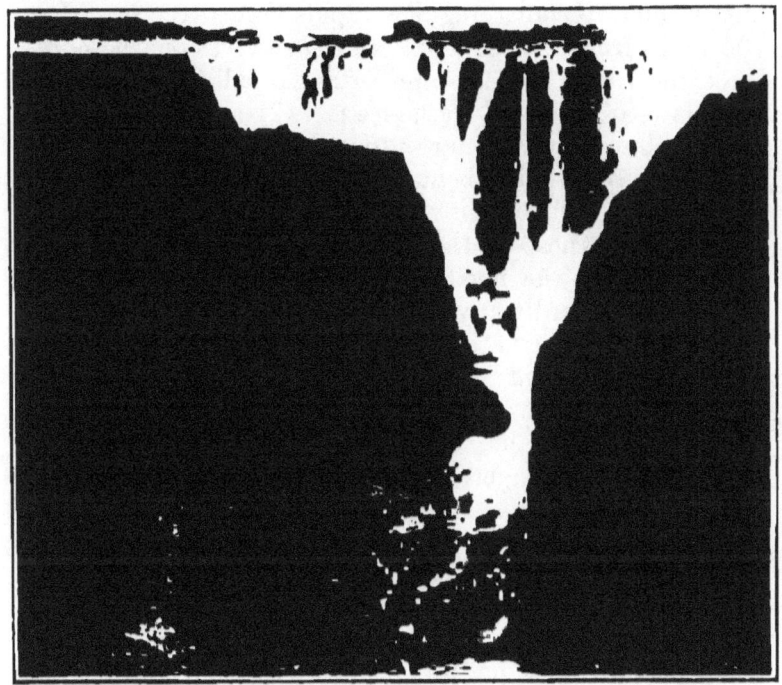

Fig. 27.—The Victoria Falls.
(Photograph by Mr. W. Ellerton Fry.)

as rise from the shores of the lakes in the great valleys of East Africa. On the south the streams which descend from the plateaux of Matabili and Mashona Lands are smaller, and many of them flow in broad sandy beds, with little water above the surface in the dry season. The two principal are the Sanyati and the Hanyani, or Panyami. They likewise have a decided westerly trend, or the reverse of that of the main river, due to the fact already mentioned that the greatest heights lie towards the east side of the continent. At about 300 miles in a direct line from the sea, navigation is again broken by dangerous rapids—those of Kebra-basa—which mark the passage through a part of the East African highlands. From this to the sea the river is generally wide, and though in parts shallow and obstructed by sand-banks, presents on the whole a good navigable waterway towards the interior. At the Lupata Narrows it again passes through a ridge of high ground, but without forming regular rapids. A little less than 100 miles from the sea the Zambesi receives on the north bank the Shiré, the most important of all its tributaries, as it flows from Lake Nyasa (see p. 147), and thus supplies a navigable waterway far into the interior, broken only by a series of rapids about midway between the lake and the Zambesi. The main river forms a delta at its mouth, smaller than those of the Nile and Niger, but still considerable. On the north side some small coast streams which receive hardly any water from the Zambesi approach very near to the true branches of the delta, and therefore increase its apparent size. The principal arms enter near the south-east end of the delta, the largest being known as the Eastern Luabo, or Koama River, but their navigation is rendered dangerous by bars, on which breakers are constantly beating. A smaller branch farther north named the Chinde (surveyed by Mr. D. J. Rankin in 1889) is more readily accessible, and has now become the main channel of entry to the Zambesi. The whole delta is flat and swampy, covered in part with mangroves, and is malarious. Life in it is made almost unendurable by myriads of mosquitoes.

The South-east Congo Basin. The south-east parts of the Congo basin are closely connected with the Zambesi region, both as occupying part of the same plateau and being most easily reached from the sea by the Zambesi route, and also on political grounds. The surface falls westwards in steps towards the Upper Congo and its lakes, and towards the north-east by a steep cliff towards the streams entering Lakes Rukwa and Nyasa. It is generally undulating, and the valleys of streams are often swampy, forming the "sponges" described by Livingstone. Between Lakes Tanganyika and Mwero there is a large salt swamp drained only intermittently towards the Congo system.

The Coast-lands.—Between Cape Delgado, just south of the Rovuma, in the north and Delagoa Bay in the south, the African coast forms two wide out-curves of unequal size, that to the north being the larger. These two curves meet in about 20° S. latitude, forming (for Africa) a fairly marked bay. Owing to the narrowing of the continent southwards, the smaller southern out-curve is considerably farther west than the northern, and the coast, as a whole, faces south-east rather than east. A certain distinction in the surface features of the coast-lands corresponding to these two curves of the coast may be observed, although in some ways the lower course of the Zambesi, which more nearly bisects the whole area, is the most important line of division. In the north the general altitude is greater, the interior plateau (with a general altitude of perhaps 1500 feet) approaching the sea and leaving quite a narrow strip of lowland along it. In the south, on the contrary, the land forms a generally level plain of no great altitude. The surface of the northern section is much broken by mountains, both in the form of ridges and isolated peaks, mostly composed of granite or other crystalline rocks. The most important groups of mountains occur in the district east of the Shiré valley not far from the south end of Lake Nyasa, the best known being Mount Zomba (7000 feet), between the Shiré and Lake Shirwa; Mount Mlanji (9600), south of Lake Shirwa; and the Namuli group, some distance east of the lake. As a rule, the separate mountain groups rise steeply from the plains on all sides, but in their upper parts have somewhat of a plateau character, though seamed with ravines and rising above the general level in single ridges or peaks.

North of the Zambesi a number of independent streams traverse the plateau in an easterly or south-easterly direction. They are generally rapid and shallow, and consequently useless for navigation. The Lurio, which flows north from the Namuli Hills before turning eastwards, has the longest course. South of the Zambesi the Pungwe has somewhat the same character, but its valley is important as affording the best inlet to the Mashonaland plateau, on which its sources lie. It enters the sea by a wide estuary almost in the angle between the two out-curves of the coast. Between it and the Lower Zambesi a marshy tract runs behind the coast mountains, forming a connecting link between the two systems. The Sabi, farther south, is much longer; it flows southwards on the upper plateau for a long distance before turning eastwards across the plains, where its bed is wide and sandy. The Limpopo drains a still larger area of the high plateau, and only in its lower course traverses the lower plains. The greater part of its basin lies to the south of its main stream, within the region to be dealt with in the next chapter. Between the Sabi and Limpopo there are few permanent streams. The country is of a very uniform level, and hardly anywhere over 800 feet above the sea. There are many hollows which form pools in the rainy season, when the beds of the water-courses also become filled. In the dry season water is often obtained with difficulty. Much of the surface is covered with dense bush, alternating with forests of timber trees.

Climate, Flora, and Fauna.—Broadly speaking, the Zambesi region is very similar to East Africa in these respects, although there is perhaps greater uniformity over the whole area, owing to the small amount of ground above 5000 feet, while by reason of its greater distance from the equator it is withdrawn from the influence of the equatorial rainfall, a larger part of the year being dry. The coast-lands, though subject to a high temperature, have therefore a less dense covering of damp tropical jungle. The greater part of the interior is either grass-land with scattered trees, the baobab being still a characteristic form, or covered with the thin East African forest. One of the largest forest-clad areas occurs on the uplands west of the Loangwa valley. On the banks of streams and on mountain slopes a more luxuriant forest occurs, but the mountain forests are not so extensive as farther north. Except in the south-west, arid steppes are not common, the most important tract of this character lying between Lake Nyasa and the Loangwa.

The animal life of the Zambesi region is similar on the whole to that of East Africa. The hippopotamus abounds in the waters of the main stream and its tributaries, and the elephant occurs almost everywhere except in the coast-lands. Towards the west and north-west, however, animal life becomes scarcer. A large part of the country, especially in the central Zambesi basin, is infested by the dreaded tsetse fly, whose bite is fatal to horses and cattle, and which thus proves (where present) a great obstacle to the opening up of the region.

Minerals.—The southern part of the Zambesi basin falls within the gold-bearing region of South Africa, and was one of the earliest known sources of gold supply. Other minerals are found in the same districts, and coal has been discovered on the Lower Zambesi and on Lake Nyasa.

2. *Inhabitants of the Zambesi Region*

The native inhabitants of the Zambesi region show a still greater uniformity than the climate, belonging almost exclusively to the Bantu branch of the Negro race. There is no large region in all Africa in which there is so little mixture of races. Among the Bantus a broad distinction may be drawn between the tribes related to the Zulus who have pushed northwards from South Africa (some within quite recent years) and the older inhabitants, who have as a rule been subjected by the invaders. The Matabili, on the plateau south of the Zambesi, only reached their present home in 1840, when they established a military empire over the Mashonas and other tribes previously settled in the country. In the southern coast-lands, too, a race of Zulu affinities has of late years held sway over the native Tonga tribes under the chiefs Umzila and Gungunyana, but the Zulu power, both here and in Matabililand, has been broken by contact with Europeans. The Zulu tribes south of the Lower Zambesi are known to the Portuguese as Landins (couriers). The Angoni, west of Lake Nyasa, represent earlier migrations of Zulus from the south, and they have mixed much with the other tribes. They live largely by plunder and have no political organisation, but are of fine physique, and those who have engaged themselves as labourers in the British settlements are praised for their honesty and steady hard work.

Of the tribes which are not Zulus, the Barotse and Mambunda on the Upper Zambesi have formed an extensive and powerful kingdom, but the Makololo, a branch of the Basutos from the south, who held sway in these parts in the middle of the nineteenth century, have now lost all power. The Yao, east of Lake Nyasa, and the Awemba, between Lakes Tanganyika and Bangweolo, are noted for their predatory habits. Arab (*i.e.* Mohammedan) influence has hardly reached the Zambesi region except in the country round Lake Nyasa. Here, however, its ill effects have been particularly marked, and the slave trade has, until quite lately, resisted all efforts to put it down.

3. *Portuguese East Africa*

History.—The political connection of the Portuguese with this coast began almost immediately after their first voyage to India. Of the old Arab marts which had flourished long years before they appeared on the scene, Sofala was taken and made tributary in 1505, and Mozambique only two years later. Other places were soon occupied, such as Quillimane, on a stream long considered one of the mouths of the Zambesi, and Sena, Tete, and Zumbo, on the lower and central course of that river, which supplied a natural route into the interior. But although missions were founded at some of these places, and traders and at least one scientific expedition (that of Dr. de Lacerda in 1798) visited the more remote interior districts, Portuguese influence never extended far from the actual posts occupied, and no footing was obtained on the comparatively healthy interior plateaux. Although traversed by Arab trade routes, the plateau between the coast and Lake Nyasa had remained quite unknown until explored by Mr. O'Neill in 1881-84. The rights of Portugal are now recognised over the whole country bordering the coast to a width (for the most part) of 200 to 300 miles, reaching to the east shore of Lake Nyasa and extending up the Zambesi as far as the Loangwa, but little has yet been done to open up these extensive territories, and the revenue of the colony falls far short of the expenditure. The authorities exercise but small control over the remoter stations, which are often in the hands of half-caste officials of the lowest character, bearing the title of "capitão mor," many of whom are descended from convicts formerly transported to the colony. The slave trade has been actively prosecuted at many of these stations until quite recently.

Government, Towns, etc.—Portuguese East Africa is divided into two provinces separated by the Zambesi, that of Mozambique in the north, and of Lourenço Marques in the south, named after their chief towns, at which the governor now resides alternately. Mozambique (long the capital of the whole territory) is built, like Mombasa, Kilwa, and other East African ports, upon a small coral islet, which gives shelter to the harbour of Mosuril Bay, on the side of the mainland, where is a suburb. As a result of the failure of the Portuguese to develop their East African possessions, the old trade centres are declining, while the towns of growing importance are those which command the routes into foreign territory in the interior. Thus the port of Chinde, at the mouth of the Chinde arm of the Zambesi delta, has arisen as the port of entry for British Central Africa, although much of the trade still passes through the old port of Quillimane. Farther south Sofala, once the most celebrated port of East Africa, exporting large quantities of gold from the interior, has fallen into complete decay, while Beira, which commands the route to Mashonaland by the Pungwe River, has risen rapidly within the last few years. The mouth of the Pungwe forms a good harbour, and river steamers can ascend nearly 40 miles from the sea to Fontesvilla, whence a railway has already been carried almost to the borders of Mashonaland. Lourenço Marques, quite at the southern extremity of the Portuguese possessions, owes its importance to its position as the natural port for the Transvaal, with which it is connected by a railway, but only 57 miles of this is within Portuguese territory. The trade of Lourenço Marques is greater than that of any other port of Portuguese East Africa. The bulk of it is carried on in British ships, which also take a large part of the trade of Mozambique.

In addition to the larger ports, there are numbers of small places along the coast where the products of the country (consisting chiefly of oil nuts and seeds, indiarubber, ivory, copra, and wax) are collected for export by agents of Indian trading firms. In the seventeenth century the Portuguese Government granted a monopoly of trade to the Banyans (natives of India), and although it has since been withdrawn, the petty trade in the products mentioned still remains entirely in their hands.

Tropical products could no doubt be grown in the colony, but the absence of a strong government stands in the way of enterprise. Sugar has, however, been produced of late years near Quillimane, and a certain amount is exported to Portugal.

4. *British Zambesia*

British influence in the Zambesi basin has been developed from two different centres. North of the main stream, the highlands adjoining Lake Nyasa and the Shiré River (approached by that river from the Zambesi mouth) have for many years been the scene of a civilising work, the outcome of the labours and appeals of Dr. Livingstone; whilst on the plateau south of the Zambesi, British influence has been extended by travellers advancing from the older colonies in the south from the time of the missionary Dr. Robert Moffat onwards. Thus, although the eventual recognition of British sovereignty over the whole northern Zambesi basin has made both regions parts of one continuous territory, they still remain distinct as regards both their administration and general outlook, and are, besides, partially separated by the Portuguese territory on the Central and Lower Zambesi.

A. British Central Africa

This name has been applied to the whole territory north of the Zambesi, but it is divided into (1) the British Central African Protectorate (sometimes known as Nyasaland), consisting of the older base of operations, and (2) a large territory farther west, much of which is still under its native chiefs, to which the title "Northern Zambesia" is sometimes given.

Soon after Dr. Livingstone's death, work was begun in Nyasaland, on the one hand by the Scottish missionaries who settled at Blantyre, on the highlands east of the Shiré, and at certain points on the lake shores, and on the other by a trading corporation (the "African Lakes Company"), the object of both alike being to free the country from the atrocities of the slave trade. In order to develop the favourable waterway into the interior supplied by the Shiré and Lakes Nyasa and Tanganyika, a road was constructed between the two lakes at the expense of Mr. James Stevenson, from whom it is known as the "Stevenson Road," another road supplying means of transit past the rapids on the Shiré, while steamers were placed on Lake Nyasa. Many difficulties were encountered both from the armed opposition of the Arabs and from the claims of the Portuguese to the sovereignty of the district; but these latter were at last disposed of by the Anglo-Portuguese agreement of 1891. The administration was soon afterwards placed in the hands of a commissioner responsible to the Crown. The protectorate comprises the Shiré basin above the Ruo (a tributary on its eastern bank), a small part of the eastern shore of Lake Nyasa, and the whole western shore as far as the Songwe River, at its north-west end. A large area to the west, embracing a part of the Congo basin as far as Lakes Tanganyika, Bangweolo, and Mwero, had been also recognised as British territory, and this was for a time under the control of the Nyasaland commissioner. However, although naturally approached by the Nyasa route, its administration has been since entrusted to the British South Africa Company.

Fig. 28.—Post Office at the Mlanji Station. (From the *Geographical Journal*. Photograph by Mr. Alfred Sharpe.)

Progress in Settlement. There is perhaps no other part of tropical Africa in which so much progress towards civilisation has been made as in Nyasaland, and this in spite of the bitter hostility of the Arabs and slave-trading chiefs. In the work of breaking their power great assistance has been rendered by gunboats launched on the lake and a force of native troops introduced from India. The whole country bordering on the route to the more distant lakes has been brought under control, and divided into sixteen districts, placed under British officials. In each of these, stations have been founded, the most distant being Rhodesia on Lake Mwero, and Fort Rosebery on the Luapula River. Regular communication with these is maintained and a fixed postal service has been instituted. Besides the gunboats, there are steamers both on the Shiré and Lake Nyasa, and a steel sailing vessel has been placed on Lake Tanganyika. The headquarters of the Commissioner are at Zomba, in the Shiré highlands. Since about 1890, the year in which the country was definitely secured for Great Britain, settlers have entered the country in considerable numbers, and the total now reaches two or three hundred, a large number for a district in the interior of tropical Africa. Most of these have taken up their abode in the Shiré highlands, where they enjoy fairly good health. Substantial houses have been built, and the settlements have almost the character of European villages. Some of the higher ground to the west and north-west of Lake Nyasa is, however, more favourable for permanent colonies of white men. Natives of India too have already settled in the country, and much is hoped from their industry and settled habits as an example to the Negro. The missionaries, who were the pioneers in all this civilising work, are still actively employed, and at Blantyre there is a handsome church, while large numbers of natives are taught in the schools. Some of the tribes have shown themselves quick at learning mechanical arts.

Resources and Trade.—Besides the development of trade in native products such as ivory and oil seeds, much has been done in the way of systematic planting, especially of coffee, which is by far the most valuable export. The amount reached 165,000 lbs. in 1894, and is rapidly increasing. Experiments are being made with other crops, and it is thought that tea and cacao will give good returns. Other products are rice, sugar, cotton, and indiarubber. Wheat might probably be grown with success on the higher plateaux, which also produce good pasturage and are apparently free from the tsetse fly. Valuable timber occurs in many parts, one of the most useful trees being the cypress (*Widdringtonia Whytei*), found only on the higher parts of Mount Mlanji. Iron is found over a large part of the surface, and coal has been discovered on the north-west shore of Lake Nyasa. Calico is the principal import, the bulk of it coming from Great Britain.

B. Matabili and Mashona Lands

Occupation.—The discovery of gold in this region by the explorers Mauch and Baines between 1865 and 1870 gave the first indication of its prospective value, and led at once to schemes for the utilisation of its resources ; but it was only in 1888-89 that British influence became supreme by the treaty concluded with the Matabili king, Lo Bengula, soon followed by the formation of the "British South Africa Company" (largely due to the initiative of Mr. Cecil Rhodes) as the administrative authority. The principal sphere of operations was the territory of the Matabili and the subject Mashonas—an area consisting of plateaux between 3000 and 5000 feet above the sea, traversed by the ridge of hilly country which divides the southern tributaries of the Zambesi from streams flowing southwards to the Limpopo and Sabi rivers. In addition to the presence of gold, the advantages of this region consist in its comparative healthiness in the higher parts and in the fertility of its plains, which are suitable both for agriculture and stock-raising.

Owing to the warlike character of the Matabili, Bulawayo, the residence of their king, was at first avoided, and the spot chosen for the beginning of operations was in Mashonaland, near Mount Hampden, at the sources of the Mazoe and Hanyani, at an elevation of nearly 5000 feet. By reason of the great distance from the settled districts of British South Africa, the first difficulty to be overcome consisted in the opening up of communications. The route from Cape Colony was fairly practicable as far as the Macloutsie River in about 22° S., in which direction the extension of one of the Cape lines of railway was at once set on foot. Northwards a road was cleared through the bush by Mr. Selous on the eastern slope of the country to the site of the destined settlement, to which the name Fort Salisbury was given. After difficulties with the Portuguese with respect to territorial claims had been adjusted, a railway was also commenced from the east by way of the Pungwe River, which affords by far the shortest line of communication with the coast. Meanwhile a road was cleared for immediate use in this direction. A rush of miners and agricultural settlers soon followed, and both Salisbury and other stations (Tuli, Victoria, and Charter) along the road from the south soon became thriving towns, with churches, schools, banks, and most of the accompaniments of civilisation. A war with Lo Bengula in 1893, brought on by his raids on the Mashonas, having been carried to a successful issue, his capital, Bulawayo (on the north-west side of the Matoppo Hills), became one of the chief European centres, and many other settlements were formed in various directions. The southern line of railway has been opened as far as Mafeking (a little north of 26°), and that from the east as far as Chimoio, near the borders of Mashonaland. An extensive system of telegraphs has been opened, connecting the chief towns both with Cape Colony and with Beira, at the mouth of the Pungwe, and a connection with Blantyre, in the Shiré highlands, is in an advanced state.

The name "Rhodesia" is often applied (unofficially) to the whole territory under the rule of the South Africa Company, including the districts north of the Zambesi.

Resources and Prospects.—Besides gold, of which extensive fields occur, chiefly along the line of the main water-parting, many other minerals have been discovered. The agricultural capabilities, especially of Mashonaland, are said to be excellent, and the higher parts beyond the range of the tsetse fly are suitable for cattle and sheep. A great check to progress has, however, been caused by the native revolt which broke out among the Matabili early in 1896, and has since spread to the Mashonas. It originated in the discontent arising from the interference of the white men with the plundering habits of the Matabili, and the hardships caused by loss of cattle from disease. The progress of the country must be retarded considerably by this rising, but with its great natural resources it may be hoped that the ground lost will be recovered in time.

CHAPTER XII

SOUTH AFRICA

1. *Physical Features*

General Characters.—Sharply defined as it is by the ocean on the east, west, and south, there is no decided boundary to the north which separates South Africa from the more northern regions already described, if we look merely at the general form and relief of the country. In these respects the two regions may be said to merge gradually with one another. The main distinction lies in the fact that the greater part of the area now to be considered lies without the tropics, and that differences of climate, reacting in turn on the growth of vegetation, bring about corresponding differences in the general conditions of life and the capabilities of the country.

The **geological formations** represented are unusually varied for Africa, and the country is, as a result, rich in a variety of minerals. South of the Orange and Vaal rivers we find a fairly regular succession of old sedimentary formations (primary and secondary) beginning from the south-east, where one of the oldest, known as the Table Mountain Sandstone, covers a large area. The most extensive of the secondary formations is that known as the Karroo, in the lower beds of which diamonds are found in great numbers. Farther north the oldest strata reappear both on the east and west (the eastern beds of Table Mountain Sandstone containing some of the most important gold-fields of the country), together with a large extent of crystalline rocks (gneiss, etc.), while the central parts are largely covered with recent sands.

Relief.—In no other part of Africa does the high interior plateau so completely fill up the whole surface as in the extreme southern end of the continent. From the mouth of the Limpopo on the east coast to that of the Kunene on the west—a distance of 2500 miles—the lowlands, under 2000 feet, form the merest fringe round the edge of the plateau, which is decidedly broken at one point only, the mouth of the Orange River. The South African plateau is also marked by great simplicity of general form, the higher ground keeping everywhere close to the outer rim, while the interior forms by comparison a vast shallow hollow, sinking, however, nowhere so low as 2000 feet above sea-level. The northern parts of this central hollow appear to have been formerly occupied by large lakes. This arrangement of the higher and lower lands gave rise to the comparison with an inverted plate or saucer, which was originally made with special reference to South Africa, though afterwards extended to the continent as a whole. The valley of the Limpopo forms in the east somewhat of a break in the outer zone of highlands, and the plateau of Mashonaland, by which that zone is continued north of the river, nowhere reaches the height generally maintained in South Africa.

The bordering highlands may be divided into two sections separated by the valley of the Orange, the one running parallel with the south-east and south coasts, the other bordering on the west coast. The former attains the greatest width and height towards the north-east, or where it runs parallel with the south-east coast. Its outer edge is here marked by the important range of the Drakensberg, whose highest peak, the Mont aux Sources, reaches 11,150 feet. Towards the interior the country falls more gradually, forming (in the Boer republics) a broad elevated tableland with an undulating surface. At the Mont aux Sources the Drakensberg divides into two branches separated by the upper valley of the Orange River. That in the direct line of the range dies out after a time, but the other branch, which bends south-east towards the coast, is continued in one of the ranges parallel to the south coast. Here there are three fairly definite parallel ridges marking successive steps from the coast towards the interior. The two southern are broken through by the coast streams and are separated by the longitudinal valleys of their tributaries, which flow for long distances parallel to the coast. The first range (from the south) rises in the Lange Berge to

FIG. 29.—MOUNTAIN SCENERY ON THE SOUTH-EAST COAST.
(Photograph by Mr. E. H. V. Melville.)

a height of 5600 feet, and the second in the Groote Zwarte Berge to almost 7000 feet. A much wider interval (named the Great Karroo, which will be spoken of later) separates the second from the third parallel range, which is the continuation of the Drakensberg, and forms the main water-parting of the country. It is known by different names in different parts, the Sneeuw Berge (rising in the Compass Berg to 7800 feet) and the Nieuwveld range (6200 feet) being among the most important sections. North of these the land slopes down gradually to the Orange River.

The highlands which border the south-west coast have rather the character of a strip of plateau, from which isolated ridges and table mountains rise, than a regular mountain chain. About the latitude of Walfish Bay it attains its greatest width, rising from the coast in a series of terraces and traversed near its inner edge by mountain ranges, which fill up the greater part of Damaraland. The highest summit is Mount Omatako (7500 feet), in about 21½° S.

The central hollow has a generally uniform surface, largely covered with sand, and hardly broken by mountains anywhere. At about the Tropic of Capricorn it is divided into two basins by a slight swelling of the surface, from which the ground sinks gradually north and south.

Rivers.—The two principal streams of South Africa, both of which flow for the greater part of their course on the interior plateau, are the Orange River and the Limpopo, the former breaking through the outer rim to the west, the latter to the east. The **Orange River** derives almost its whole supply of water from the Drakensberg range, in which the sources of two main branches lie. The south branch, although the shorter, is generally considered the principal stream. Its upper valley lies between the two main branches of the Drakensberg, dominated by the highest peaks of the range, and as it flows for a long distance generally parallel with the mountains, it receives a larger supply of moisture from them than the northern branch (the Vaal), which passes more quickly out of their influence. After the junction of the two branches the river passes through an arid region, and is joined by no more permanent streams, although a large area, especially to the north, falls naturally within its basin, and is furrowed by channels which contain water at times. At about 350 miles from the sea the Orange precipitates itself by many a fall and rapid down a gorge 16 miles long and 300 feet deep. These falls are known as the Great Anghrabies, or George IV. Falls. Before the sea is reached rapids again occur, and the mouth is blocked by a dangerous bar, so that the river is almost useless for navigation. Its volume is great when the rains occur in the mountains, but at other times it has little water.

The **Limpopo**, called sometimes the Crocodile River, begins its course by flowing, like the Orange, down the inner slope of the eastern highlands, its head stream rising in the Witwatersrand, a ridge of high ground which separates the Limpopo and Orange basins. But instead of continuing its course to the west coast, the Limpopo makes a wide curve round the northern border of the Transvaal highlands, and breaking through their eastern rim, flows for some distance across the plains within the limits of Portuguese East Africa before reaching the sea north of Delagoa Bay. Most of its tributaries drain the plateau included within its curve, and the principal of them, the Olifant, has to break through both the main border range and a second parallel ridge, known as the Lebombo range, before joining the main river. The Limpopo is generally shallow and rapid, but can be navigated in its lower course by light draught steamers during a portion of the year.

The northern interior of South Africa is occupied by the basin of **Lake Ngami**, a small lake, probably the remnant of a much larger one which existed in former times. It is fed by a stream which rises far north in the moister regions of Central Africa, but diminishes in volume as it approaches the arid region of the Kalahari. It is known successively as the Kubango, Okavango, and Tioghe. When unusually full Lake Ngami overflows by the Zuga or Botletli River into a region of salt swamps, which forms the lowest depression of this part of Africa.

The coast streams of South Africa have all short courses, and have little economic importance. Those on the south-east coast carry down a large body of water when flooded by rains among the mountains, and have for the most part cut for themselves deep channels. The principal are the Great Fish River and the Tugela. In the south the streams are less copious, and on the south-west coast there are no permanent streams at all. The principal temporary watercourses are the Swakop and the Kuisip, which enter Walfish Bay. Water is generally found beneath their sandy beds.

Climate.—Although the greater part of South Africa lies within the southern temperate zone, there is a remarkable variety in its climate, and the contrasts are to be observed rather between the east and west than between the north and south of the area. Both temperature and rainfall are, broadly speaking, greater in the east than in the west, and the eastern region has more in common with the tropical regions which adjoin it on the north, showing also smaller divergencies between the seasons. This is due principally to the influence of the warm Mozambique current, which flows southwards down the coast. Although, of course, the temperature varies everywhere at different times of the year (the hottest month being January, and the coolest July), certain laws hold generally as regards its distribution over the surface. The lower temperature is experienced generally on the west coast and on the highlands which bound the basin of the Orange River southwards. The low temperature on the west coast is due in great measure to cold water rising from the ocean depths, which gives rise to dense mists. The southern highlands, owing to their elevation, have the lowest temperature of all, the mean for the year sinking as low as 53° towards their western end. They are also exposed to the greatest variations between summer and winter. The warmer zone encircles the north-eastern end of the highlands, embracing both the south-east coast (where the annual mean rises to over 70° in Zululand) and the interior basin on the north-west of the highest ground. The south coast enjoys a mild and equable climate, being neither cold in winter nor hot in summer.

The rainfall of South Africa is on the whole scanty, a fact which, though somewhat unfavourable to plant growth, makes the climate particularly healthy and especially beneficial to those suffering from lung disease. In the east the larger amount of rain is due to the influence of the Drakensberg range, which condenses the moisture from the south east winds, and acts as a screen to keep it from the regions eastwards. Although the west is generally dry, there is a small rainy region in the extreme south-west in the neighbourhood of Cape Town. But whereas the eastern region gets most of its rain in summer when the moist winds blow towards the heated land-surface from the South Indian Ocean, in the south-west the rain-bearing winds are those which blow in winter from the north-west towards the area of low pressure in the Southern Ocean, and the rainy and dry seasons are exactly reversed. In the intermediate region on the south coast rain is more evenly distributed through the year, but the Karroo region north of the two southern steps of the plateau receives very little rain at all. In the whole eastern section north of the Orange River the only district with any considerable rainfall is that of the Damara highlands, where, at the most, 20 inches fall, chiefly in the early summer.

Vegetation.—Corresponding with these variations of temperature and rainfall there are great differences in the vegetation of the country, although certain characteristics hold good over a large part of the surface, being the result of the dryness. These are a deficiency in trees and a general stunted growth of plants and sombre shade of foliage, but as a compensation for this flowers are usually brilliant and varied. The heaths form the most characteristic family of plants, and their variety in South Africa is greater than in any other country. The absence of the baobab tree and (except on the south-east coast) of palms serves as a broad distinction between South Africa and the tropical African savannahs. The whole of South Africa may be broadly divided in respect of its vegetation into two regions, each of which again may be subdivided into smaller areas according to their more special features.

(1) The lands bordering on the south-east and south coasts are covered in part with scrub or forest. Towards the north-east the vegetation of this zone is sub-tropical, owing to the heat and moisture, having many resemblances to the vegetation of the more northern parts of East Africa, and containing a single representative of the family of palms, *Phoenix reclinata*. On the west the plants belong rather to the temperate zone.

(2) The rest of South Africa is as a whole composed of more open country, but three separate areas may be

distinguished : (*a*) the Karroo, a name more particularly applied to the space intervening between the second and third ranges parallel with the south coast, but which may be extended so as to include the whole country northwards almost to the Orange River and westwards to the sea. It is generally a dry elevated moorland covered with stunted bushes of heath or other shrubs, with little grass. The surface becomes baked and hard, and allows the scanty rain which falls in summer to run quickly away instead of sinking into the soil. (*b*) The Kalahari, occupying the whole western half of South Africa north of the Karroo. Its surface is generally sandy, with very little visible water, but the rain which falls sinks below the surface, and a supply can mostly be reached by digging. With the exception of the highlands running parallel to the sea, the surface is generally a level grassy expanse, though some parts form sandy or stony wastes, especially near the coast, and towards the north there is much more bush, and even some forests. (*c*) The eastern highlands, occupied chiefly by the Boer republics, are known as the "Veld" region. Here again most of the surface is grassy, but the soil is less sandy, and the more plentiful supply of moisture gives a somewhat greater luxuriance to the vegetation.

The coast-lands which form the first main division of South Africa are mostly covered with stunted trees and bush, the primeval forests which once clothed much of the surface having been in great measure destroyed. They are still to be found in certain districts, especially on the seaward face of the mountain ranges. Near the south coast they grow down to the sea-level, and on the Drakensberg range ascend to 6000 feet. These South African forests are described as of great beauty, the tall and heavy timber trees being set off by creepers and tree ferns and interspersed with gay flowers. The most important trees are the yellow wood, a conifer known scientifically as *Podocarpus* (the most abundant of all); the stinkwood or laurelwood (*Orotea*, belonging to the family of the true laurels), very durable, and of a beautiful mottled appearance; and the sneezewood and ironwood, of which the heart-wood is hard and almost imperishable. Other kinds of trees have been introduced into South Africa, especially in the peninsula of the Cape of Good Hope.

In the steppe regions of South Africa the grasses and other plants grow as a rule in tufts, instead of forming a continuous carpet as in England. A result is that the soil is very easily washed from the surface, deep gullies being very rapidly formed where once a small furrow is made by the feet of animals or any other cause. The generally slight covering of surface soil is in its turn one of the chief reasons for the stunted character of the vegetation of South Africa as a whole, and also for the smallness of the area suitable for agriculture.

Fauna.—The wild animals of South Africa are in the main those which characterise the drier parts of the continent within the tropics, from which there is much less difference in this respect than in respect of the vegetation. When Europeans first began to settle in the country the number of wild animals was extraordinary. Elephants, the rhinoceros, hippopotamus, and vast herds of antelopes, zebras, quaggas, giraffes, and buffaloes were everywhere found, together with the lion, leopard, hyaena, and other beasts of prey. But with the settlement of white men and the relentless persecution of sportsmen, animal life has become less and less abundant, and few of the larger wild beasts are now found south of the Orange River, while many have altogether retired far into the northern interior. Elephants and buffaloes, however, still remain in one or two spots, such as the forests on the south coast. Even the antelopes are now comparatively rare in the southern parts, though springbuck (the most characteristic form), for which a close time has been imposed by Government, are fairly plentiful in the Karroo. The eland, the largest of the antelope family, sometimes 6 feet high at the withers, has been domesticated in Cape Colony, and the same is the case with the ostrich, which in its wild state has retired to the most inhospitable districts. Other domestic animals will be spoken of under the separate countries of South Africa.

2. Inhabitants of South Africa

In South Africa we have to do not merely with the native races—here belonging to three distinct subdivisions of the black type of mankind—but with a large population of white immigrants, themselves derived from various European nations, who have found here a favourable home owing to the temperate climate with which the country is favoured. This great mixture of races, and the difficulty arising from their conflicting aims and interests, form the greatest obstacle to a rapid advance in prosperity.

Native Races.—The broad distinction between the east and west which we have noticed in respect of climate holds good also as regards the native races. The more humid eastern half is occupied by Bantu tribes, while the western half is peopled principally by the Hottentots and Bushmen (spoken of at p. 23). There seems no doubt that these tribes once occupied a larger area, and that they have been driven back even within comparatively recent times by the encroachments of the Bantus, who have pushed southwards from their earlier home in East Africa. The climatic and other resemblances between South-East Africa and more northern regions helps to explain the southern extension of the Bantu in these parts, the arid regions where the Hottentots and Bushmen have held their own offering much less attraction to the newcomers.

The **Bantus** themselves may be divided into two groups lying on the whole east and west of each other, viz. the Kaffirs and Zulus in the east, and the Bechuanas and Basutos farther west in the more central parts of the continent. The Kaffirs[1] and Zulus are both noted for their fine physique. They live principally by cattle-rearing, and have been much given to cattle raids on their neighbours, which have brought them into collision with the white

[1] The name Kaffir means infidel, and was originally given to these tribes by the Arabs of East Africa to distinguish them from the followers of Mohammed. The same term has been applied to a race in Asia for a similar reason.

settlers. The Kaffirs are split up into various separate tribes which have not combined to form a political power. The Zulus, on the contrary (who live north and east of the Kaffirs), formed a powerful kingdom under Chaka early in the nineteenth century, which retained its importance down to the war with Great Britain in 1879. The Bechuanas are now split up into several separate tribes, each under its own chief. They show more inclination to adopt civilised modes of life than most of the native peoples of Africa. They have more than once come into collision with the European colonists, but have of late years shown themselves peaceable and law-abiding, and have lent their aid to the British authorities against their old enemies the Matabili. The Basutos occupy part of the mountainous district of the south-east bordering on the domain of the Zulus and Kaffirs. All these Bantu tribes practise both cattle-rearing and agriculture. The northward migrations of Bantu tribes in South Africa within the nineteenth century have already been touched upon in the last chapter.

Whilst the Bantu tribes have in great measure held their own side by side with the white settlers, the **Hottentots and Bushmen** have either declined in number or retreated before the intruders. Almost the only group of pure Hottentots remaining is that of the Namas in Namaqualand, principally north of the Orange River. Elsewhere they have mixed with other races, and adopted a semi-civilisation in the more settled parts. The chief occupation of the true Hottentots is sheep-farming, and they live under the patriarchal rule induced by wandering in search of pasture. The Bushmen are also nomadic, but live entirely on the proceeds of the chase or wild roots found in the steppes. The white settlers—of British, Dutch, French, and German origin—will be spoken of under the separate political units of South Africa.

3. *Colony of the Cape of Good Hope*

History.—The first European settlement was founded by the Dutch on the site of Cape Town in 1652, for the

Portuguese never established themselves in the country whose coasts they were the first to discover. The Dutch settlement gradually expanded into a small colony of farmers, whose numbers were increased by an influx of French Huguenots after the revocation of the Edict of Nantes (1685); but during the Napoleonic wars, Holland having become subject to the French, the Cape was occupied temporarily by Great Britain from 1795 to 1802, and recaptured in 1806, when war had again broken out after the Peace of Amiens. The Cape was finally recognised as a possession of Great Britain by the Treaty of Paris, concluded in 1814. At this time the Dutch and Huguenot settlers, who had become united into a Dutch-speaking community by the prohibition of the use of any but the Dutch language in the schools, numbered between 20,000 and 30,000, confined almost entirely to the south-west corner of South Africa. Farming operations were carried on by the help of Negro slaves introduced for that purpose. Although since the British occupation the new settlers who have arrived from Europe have been chiefly British, the older settlers have prospered and multiplied, so that Dutch and English now dwell side by side in the colony without amalgamating into a single race. It is this fact which has introduced an element of uncertainty, which attaches to the future of the country.

The abolition of slavery in 1833 caused great discontent among the Dutch, and to it was largely due the first great movement of the Boers (as the Dutch farmers are called) into the more remote parts of the country. This movement is spoken of as the great "trek," the Dutch word for travelling by waggon, this being the general means of locomotion in South Africa wherever railways have not been introduced. The waggons are drawn by long teams of oxen, numbering sometimes a dozen yoke. In this way Boer states have been founded beyond the limits of Cape Colony. The northward movement has taken place chiefly in the eastern districts. In the western parts of the colony a large Dutch section has remained down to the present, and forms even more than half of the white population.

Extent and Population.—The area of the colony

has steadily increased since the British occupation. In the course of expansion repeated collisions arose between the settlers and the warlike Kaffirs and Basutos in the south-east, but the result of several wars with these tribes was to bring their territory under British jurisdiction. Natal, first settled by the Boers but afterwards annexed to the Cape, became a separate colony in 1856. Various districts between Natal and Cape Colony long remained under native chiefs, but in 1894 the last remaining native area (a part of Pondoland) became incorporated with the British possessions. In the north the Orange River long formed the boundary, though hunters and explorers penetrated into the wilds beyond. But in 1871 Griqualand West became British territory, being annexed to the Cape Colony in 1877; and in 1895 a part of Bechuanaland, which had since 1884 been directly under the rule of the Crown, was also taken over. The population of Cape Colony (without Bechuanaland) was in 1891 a little over a million and a half, of which total the whites numbered just under one quarter.

Government.—After the British occupation the colony was at first placed under the rule of a Governor, assisted after 1825 by an executive council. Parliamentary government was first introduced in 1853, and took its present form in 1872 by an Act providing for "Responsible Government." The Governor is still appointed by the Crown, and is assisted by a ministry of six members, but legislation rests in the hands of the two bodies elected by the colonists, known as the Legislative Council and the House of Assembly.

Resources, Industries, etc.—The prosperity of Cape Colony depends principally on its farming and mining industries. The great difficulty which farmers have to contend against over a large part of the country is the prolonged drought during a part of the year. This renders irrigation necessary. Several systems have been introduced and large dams have been constructed in many places. Both of the main branches of **farming**, stock-rearing and agriculture, are represented, but owing to the general deficiency in deep rich soil, stock-rearing is by far the most important.

According to the varying capabilities of different parts

of the colony, the attention of the farmers is devoted to different classes of stock. Thus sheep form the greatest proportion in the arid lands between the Orange River and the southern mountains. Cattle, as well as horses, etc., are reared in largest numbers on the coast-lands both in the south and south-west, and goats are in larger proportion a little farther inland. Some numbers both of cattle, horses, and goats are reared also in Griqualand West north of the Orange River, but very few sheep. Ostrich-farming, for the sake of the feathers, has come into importance during recent years (the wild ostrich having retired beyond the Orange River), and is practised everywhere, but principally in the south and south-east. Large "camps" enclosed by wire or other fences are formed for the birds, which feed on the Karroo bushes, and in times of drought are supplied with lucerne or other green crops. Returns made in May 1895 gave their total number as 253,000. Sheep were originally bred chiefly for their wool, which has long been one of the principal exports, the native hairy breed having been crossed with the merino; lately, however, a great demand has sprung up for mutton for the mining centres of the north. The chief difficulties in the way of sheep-farming are on the one hand the ravages of jackals and other wild beasts, and on the other the attacks of certain parasites and diseases. The most useful variety of goat is that known as the Angora goat, from its original home in Asia Minor. Its wool likewise forms an important export. Dairy-farming has made satisfactory progress of late years.

Agriculture is practised in various parts of the country, excluding the Karroo region. Wheat is the cereal most grown, and is followed by Indian corn (mealies), oats, barley, and durra or "Kaffir corn"; other important crops are potatoes and tobacco. In the south-west near Cape Town the vine is largely grown and much wine is exported. The climate is eminently suited for fruit-growing, but this industry has not yet been developed as it might be.

Of the **mining industries** that of diamonds is by far the most important. The diamond fields lie north of the Orange River in Griqualand West, occupying the lower beds of the Karroo formation known as the Kimberley Shales. The first stones were discovered (south of the Orange River) in 1867, and the rush of miners to the site of the present mines began three years later. Since that time there has been a vast development of the industry, so that the export of diamonds now exceeds in money value that of any other product of the colony, amounting to about £4,000,000 worth each year. Important copper mines are worked in the barren district of Little Namaqualand, south of the Orange River, those of Ookiep being the principal. Coal occurs in the divisions of Albert and Wodehouse, in the eastern part of the colony, and other minerals which exist, but are not largely worked, are lead, iron, manganese, etc.

The foreign trade of Cape Colony is carried on principally with Great Britain, and the chief exports are diamonds, wool, hides, ostrich feathers, copper ore, and wine. Cotton and woollen goods, hardware, and various other articles are supplied in return by the mother country.

Towns.—As the bulk of the population is occupied with farming, the towns are of no great size, and are scattered at wide intervals over the country. Excluding the more recently-annexed native territories east of the Kei River, the towns are most numerous in the south and east, where the rainfall is greatest, and where agriculture as well as stock-rearing can be practised with good returns. The whole of the Karroo region occupying the north and north-west is sparsely peopled, and the towns are both small and few in number. The principal towns may be divided into two groups, corresponding to the two areas in which the earlier colonisation was most active, and to which the two principal ports of the colony, Cape Town and Port Elizabeth, supply the bases of communication with the rest of the world. These are, in fact, the only towns with good natural harbours in the whole colony, for the coast is generally deficient in sheltered bays. Saldania Bay, on the west coast (not to be confounded with the Saldania Bay of the old navigators, which was the modern Table Bay), is an exception to this rule, but, owing to scarcity of fresh water in its neighbourhood, its shores are almost deserted.

Fig. 30. Cape Town and Table Mountain.

Many of the **western towns** were founded by the Dutch before the British occupation, but they include others which have been settled since by British colonists. Cape Town (population 51,000), on Table Bay, at the northern side of the peninsula which runs down to the Cape of Good Hope, has a twofold importance, firstly, as the capital of the colony and the port by which a great part of its trade is carried on, and, secondly, as one of the most noted stations for coaling, etc., on the ocean routes of commerce in the southern hemisphere. Its importance as a port of call on the sea route from Europe to the east led to its original occupation by the Dutch, and it is the same reason which at the present day makes its possession of vital moment to Great Britain, owing to the great extent of her possessions in the East. To Australia the Cape route is not very much longer than that by the Suez Canal, while to India it is still the only route used by sailing ships. The town lies between the shore of Table Bay and the abrupt northern cliffs of the Table Mountain, so named from its almost level top, on which clouds often rest. The background is completed by two peaks, which rise on either side, a little in advance of the Table Mountain, that to the west being known as the Lion Mountain, and that to the east as the Devil's Peak. The town thus occupies a sort of amphitheatre, rising gradually over the lower slopes of the hills. The docks and harbour lie under the shelter of the Lion Mountain, supplemented by extensive artificial works and protected by forts. There are two cathedrals, a university, a public library and museum, and a famous observatory. The streets are laid out regularly, and the suburbs are beautified by parks and gardens. The naval station of Simon's Town lies on the opposite side of the Cape Peninsula, on Simon's Bay, a minor indentation of the larger False Bay. A railway leads round the eastern side of Table Mountain, passing through a fertile district dotted with agricultural settlements, and connecting Cape Town with Simon's Bay. Port Nolloth (1500), on the west coast, south of the Orange River, is the outlet for the copper mines of Ookiep, but it is merely an open roadstead. Of the old Dutch towns which are scattered over the strip of country bordering the south coast, the most important is Paarl (7000), which is not equalled in size by any of the newer British settlements in the western districts.

The Eastern Towns.—The eastern districts of Cape Colony began to be settled soon after the final cession of the country to Great Britain, 4000 English farmers having

FIG. 31.—BOTANICAL GARDENS, CAPE TOWN.

landed at Port Elizabeth in 1820. This town (23,000) is now the chief commercial outlet for the eastern part of the colony. It lies on the western shore of the spacious Algoa Bay, sheltered on the west by Cape Recife. The port is, however, exposed to the south-easterly gales, which sometimes blow with much violence. In spite of this drawback, the trade has largely increased of late, and now exceeds that of Cape Town. The town has several fine buildings. Other ports are the newly-established Port Alfred, at the mouth of the small river Kowie, 80 miles east of Port Elizabeth, and East London, quite at the extremity of the settled districts, but neither has yet a good harbour. The principal inland settlement is Grahamstown (also dating back to the early years of the century), which is the administrative centre for the whole eastern district. It has woollen manufactures and is the seat of bishops both of the English and Roman Catholic churches. It lies about thirty miles inland from Port Alfred. There are many other small but thriving market towns in the eastern districts, most of them on one or another of the lines of railway which run inland from the ports, and which will be spoken of presently. Graaf Reinet deserves mention as the farthest town founded by the Dutch in this direction. It lies near the eastern end of the Great Karroo, where the increasing rainfall admits of agricultural operations.

Kimberley, in the angle between the Orange and Vaal rivers, owes its origin entirely to the diamond mines in its vicinity. Although before 1870 its site was a bare and arid plain, its population was in 1891 over 28,000. Its streets are lighted by electricity, and a water supply has been laid on from the Vaal River. The mines were formerly open workings, but now consist of four principal shafts or "pipes," which have been sunk to a great depth. They are worked largely by native labour, and are enclosed by high fences to prevent the abstraction of the stones by the workers. The stations of Vryburg and Mafeking, north of Kimberley, are in the part of Bechuanaland now administered by Cape Colony.

Railways and Roads.—The railway system of Cape Colony is very simple. The trunk lines start from the principal ports and make for the interior, the point specially aimed at being the mining centre of Kimberley, which, by their means, receives food supplies from the farming districts of the south. They also supply the means of transporting the produce of the interior districts to the coast. That from Cape Town, after abruptly bending northwards and again southwards to make use of a gap in the mountains, passes in a north-easterly direction across the Great Karroo between the second and third of the parallel southern ranges, ascending to the higher plateau after passing south of the Nieuwveld. Except

Paarl, on the outer side of the highlands, it passes through few towns of any importance, and even seems to avoid some of the small centres which lie near its route. It crosses the Orange River a little above Hopetown. Beyond Kimberley it has been continued to Vryburg and Mafeking, and is being pushed farther towards Mashonaland. The other trunk line has termini both at Port Elizabeth and at Port Alfred, but at Alicedale junction the two branches unite (that from Port Alfred having passed through Grahamstown) and the line ascends the valley of the Great Fish River, joining the Cape Town line some distance south of the Orange River. Before this it sends a branch north to the Orange Free State and Transvaal. Another line from Port Elizabeth passes Uitenhage and ends at Graaf Reinet, while still another starts from East London and proceeds north-west towards the Orange Free State, by Cathcart, Queenstown, and Molteno, passing, in the neighbourhood of the last-named, through the chief coal-fields of the colony.

No railway as yet supplies communication between the east and west in a direction parallel to the south coast, but this depends still on the roads, of which a large mileage is kept in good repair.

4. *Natal and Zululand*

The name Natal dates from the first discovery of the coast by Vasco da Gama, which happened on Christmas Day 1497. Early attempts at settlement, both of Dutch and British, led to no important result, until in 1837 the discontented Boers from Cape Colony founded Pietermaritzburg as the capital of the new "republic of Natalia." The next year, however, British supremacy was established, many of the Boers submitting to it, and after forming part of Cape Colony for a time, Natal was formed into a separate British colony in 1856. In 1893 a constitution similar to that of Cape Colony was granted, by which self-government was secured. The colony stretches inland from the coast to the crest of the Drakensberg range, occupying a narrow strip of lowland together with the outer terraces of the interior highlands. In 1891 the population was a little over half a million, less than one-tenth being Europeans.

The peace of the colony was often disturbed by the near neighbourhood of the warlike Zulus on the north, until the power of their King Cetewayo (who had first inflicted a severe defeat on the British army) was crushed in 1879. After a period of internal disorder and anarchy, during which the Boers of the Transvaal made themselves masters of a large part of the country, the remaining portion of Zululand was made a British protectorate (under the Governor of Natal) in 1887, and this was extended in 1895 over the intervening districts (Tongaland) up to the Portuguese territory.

Resources.—Owing to the higher temperature and larger amount of rainfall, combined with a more fertile soil than that of Cape Colony, the vegetable products of Natal are more varied than those of the sister colony. They include sub-tropical crops, such as sugar, tea, coffee, tobacco, indigo, arrowroot, etc., grown on the lower region near the coast, together with cereals and other crops of temperate regions on the higher terraces. Sugar is grown in large enough quantities to allow of some export to Great Britain, besides the supply of all the South African colonies. For the cultivation of the sugar-cane, coolies have been introduced in considerable numbers from India. On the higher grounds there is excellent pasturage, and sheep, cattle, and ostriches are reared as in Cape Colony. The mineral wealth is also considerable, the coal deposits in particular being of great importance.

Towns, Railways.—Natal has but one port, and is traversed by a single trunk line of railway. Port Natal is the best harbour on the south-east coast of Africa, and its trade, over half of which is with the mother country, is considerable. It has much increased since the opening of the gold mines in the Transvaal, but since 1889 has shown a diminution, owing to the opening of other routes from the gold-fields. The town (17,000), which is placed on the inner side of the port, is named Durban, after a former governor. The centre of Government is Pietermaritzburg (12,000), which lies at a height of 2000 feet, directly inland from Durban, at a distance of 40 miles. The main line of railway starts from Durban, and, passing through Pietermaritzburg, crosses a spur of high ground into the upper basin of the Tugela. After crossing the river at the small town of Colenso, it reaches Ladismith, and there divides into two branches. One of these runs nearly north by Newcastle, parallel with the mountains, finally crossing them, as their height decreases, near Majuba Hill, where a British force was defeated by the Boers in 1881. It has lately been continued within the Transvaal so as to join the lines from the south near the mining centre of Johannesburg. The other branch leads westwards by the Van Reenens Pass over the Drakensberg to Harrismith in the Orange Free State, and is being continued westwards to join the main line which traverses that state from south to north. There are two small branch railways from Durban parallel with the coast. The main line passes through the principal agricultural centres of the colony, and away from it. Greytown and Weenen in the north, and Richmond in the south, are the only towns deserving mention.

5. *Bechuanaland*

The central, generally arid, plains of South Africa inhabited by the Bechuana tribes have been formed into a British protectorate,—with an area of 117,000 square miles, or more than half that of Cape Colony,—part of which has been placed under the administration of the British South Africa Company, and part is still governed by its native chiefs, of whom Khama is the most enlightened and powerful. His territory adjoins Matabililand on the south-west. The Governor of the Cape Colony, in his capacity of Imperial Commissioner, exercises a general control over the country. It has not yet been settled by Europeans, except on the route to the more fertile regions of the north which skirts the Transvaal frontier; but as water is mostly present below the surface, it is suited for the rearing of cattle, goats, and sheep.

6. The Boer Republics

The two Boer republics occupy the elevated region within the Drakensberg range characterised by an undulating surface, broken by occasional ridges and isolated hills called Kopjes, and a general scarcity of trees, so that it is eminently a pastoral country. They owe their origin to the northward movement of the Boers, which followed on the abolition of slavery in Cape Colony in 1833.

The Orange Free State.—The country between the Orange and Vaal rivers was the first reached, and a new state was founded here in 1836, though its independence was not recognised by Great Britain until 1854. Of a total population (at the present day) of somewhat over 200,000 nearly 80,000 are whites. Most of these belong to the Dutch Episcopalian Church, but the English language is rapidly coming into use, and English money, weights, and measures are employed. The head of the executive is a President elected for five years, and there is a popular assembly named the Volksraad. The principal industry is the rearing of stock, especially sheep and cattle, and in smaller numbers goats, horses, and ostriches. The mineral wealth is considerable, but has not yet occasioned such a rush of miners as has occurred in neighbouring parts of South Africa. Diamonds are found in the south-west in beds analogous to the Kimberley shales, and coal principally in the north. Other minerals are gold and iron. The chief exports are wool, hides, diamonds, and ostrich feathers, the value of the diamonds exported in a single year having exceeded £400,000. There is communication by railway with the eastern ports of Cape Colony and with Durban, the main line traversing the country from south to north and passing on into the Transvaal. The capital, Bloemfontein (about 3400) is on this line, and lies in a wide plain at a height of 4500 feet. The other towns are small centres with less than 2000 inhabitants.

The South African Republic, formerly called the Transvaal Republic, occupies the space between the Vaal and Limpopo rivers. Its founders were the Boers, who, after a temporary sojourn in Natal, crossed the Drakensberg between 1838 and 1840, in order to escape from British authority. The independence of the republic was recognised in 1854, but in 1877, the country being in a state of bankruptcy, Great Britain took possession of it. In 1880 the Boers revolted, and after defeating a British force at Majuba Hill, regained the control of their internal affairs, remaining under British influence only in respect of their dealings with foreign countries. The government is in the hands of a President elected for five years, and there is a parliament of two chambers. Of late years the Boers have overrun a part of Zululand, and Swaziland (on their eastern border) was handed over to them in 1894.

The country is divided into two natural sections, the southern strip being known as the Hooge (High) Veld (the most elevated and open part), and the northern parts as the Bosch Veld, where bush and trees are more plentiful. Near the Limpopo almost tropical conditions prevail. The sole industry of the republic long consisted of pastoral pursuits (sheep being reared principally on the Hooge Veld, and cattle in the adjoining zone to the north), and there were few towns of any size, the Boers as a rule dwelling on isolated farms. Pretoria, the capital, was founded about 1855. It lies in a wide valley on the south side of the Magalies range, near the head-streams of the Limpopo. Potchefstroom, on a small northern feeder of the Vaal, is a still older town, and was for some time the chief settlement in the country.

These older towns have quite recently been eclipsed by others which have sprung up as a result of the rush to the gold-fields. Gold was first discovered in 1869, but was not extensively mined until after 1882. The most productive fields (discovered in 1885) are those of the Witwatersrand—the hilly ridge which separates the basins of the Orange and Limpopo. Here the town of Johannesburg has already attained a population of 50,000 Europeans. It is placed on a bare upland 5600 feet above the sea. Other important gold-fields are those in the district of Lydenburg in the east (known as the De Kaap fields), where the modern town of Barberton has sprung up. The miners who have flocked to these fields have been largely British, and where most numerous they already outnumber the Boers, but have not been granted the privileges of citizenship, so that a somewhat bitter feeling against the authorities has arisen, resulting in the disturbances of January 1896. Many other minerals, including coal, exist in the country.

Railways have been only lately introduced, but there are now lines from Lourenço Marques, on Delagoa Bay (p. 174), which practically serves as the chief port of the country, from Natal, and from Cape Colony *via* the Orange Free State, all converging on Pretoria and Johannesburg. The De Kaap gold-fields are reached by a branch from the Delagoa Bay line, and the construction of a line through the northern parts of the country has been decided on. The chief exports are gold and the products of stock-rearing (wool, hides, and cattle). The production of gold has reached a very high value, and is still rapidly increasing. In 1894 it exceeded £7,000,000 worth.

7. *German South-West Africa*

Although the south-west coast of Africa had long been frequented by traders and others from Cape Colony, and was looked upon as to some extent forming an appendage to that territory, no definite act of annexation applying to this region as a whole had been performed when in 1883 the attention of Germany was directed to it as a field for German enterprise. After diplomatic correspondence on the subject, Great Britain at last recognised the claim of Germany to the whole coast from the Orange River to the Portuguese boundary at the Kunene, with the exception of Walfish Bay and some of the small islands off the mainland, where the deposits of guano were worked by British subjects. Subsequent agreements fixed the interior limit at 20° east towards the south and 21° farther north, with a narrow extension at the north-east corner giving access along the Chobe to the Zambesi.

The country, which is inhabited by the Nama Hottentots in Great Namaqualand towards the south, and by the Bantu Damaras towards the north, suffers generally from scarcity of water, which limits its agricultural capabilities, though in the hilly districts of Damaraland there is rain enough during the summer months to allow of some cultivation. Cattle-rearing, however, is the industry from which the settlers have most to hope. A great advantage to European colonists is the healthiness of the climate, due to its dryness, which makes the heat supportable even in the arid plains between the mountains and the coast. Here there is a great difference in temperature between the day and night, the latter being always pleasantly cool. The interior hilly districts are free alike from great heat in summer and cold in winter. The country is rich in copper, but it has not yet been worked profitably.

Stations, Communications.—German missionaries have laboured in the country for many years, and their stations are dotted over the land. Otyimbingue, on the Swakop, or Tsoakhaub, is the centre of administration. Hitherto the outlet for Damaraland has been Walfish Bay (still a British possession), which is the best port on the coast; but attempts are being now made (1896) to establish a port in German territory at the mouth of the Swakop (Tsoakhaubmund). Communication is carried on by waggons drawn by oxen, but a project for the construction of a railway has lately been formed. Sandwich Haven and Angra Pequena Bay are the best harbours in German territory, but the country in their neighbourhood is unpromising. Colonies of Boers, who have "trekked" from the south-east, and have now accepted German authority, are found in the north of Damaraland.

CHAPTER XIII

WEST CENTRAL AFRICA

1. *Physical Features*

General Characters.—The countries with which we have to do in this chapter belong in great part to one of the best-marked natural divisions of the continent, which, however, extends also along the northern shores of the Gulf of Guinea, in the region already described in the seventh chapter. This may be described in broad terms as the region of dense forests, subject to a moist hot climate, showing small variations between the seasons, and the special home of the anthropoid apes and of the oil palm. Towards the south, however, we gradually pass into a different region with more resemblances to other parts of tropical Africa. The coast-lands are sometimes known by the general term Lower Guinea.

Relief.—The most marked feature of this region is the wide circular basin, bisected by the equator, which occupies the centre of the area, and is separated from the sea on the west by an almost continuous band of higher ground. This depression, which must in very ancient times have been occupied by a vast inland sea, but is now drained by the Congo and its tributaries, forms the only important break in the general level of the Central and South African plateau. Whereas elsewhere ground below 2000 feet above the sea is rarely found away from a narrow strip bordering the coast, in the Central Congo basin the land hardly reaches this height over an area 800 or 900 miles each way, and in the lowest parts sinks to about 1000 feet, a height less than is found elsewhere within the outer girdle of high ground in the southern half of the continent. This central basin is largely covered with recent sandstones and alluvial soil, and is the part of Africa which bears the greatest resemblance to the great plain of the Amazon in South America.

This depressed area is bounded on all sides by higher ground, which, however, forms towards the north but a slight line of elevation separating the Congo basin from that of Lake Chad. In the south there is a much longer and generally gradual slope towards the deepest hollow from the water-parting on the side of the Zambesi basin, whilst on the east the fall is fairly abrupt from the high East African plateau. The western limit is formed by a band of highlands composed largely of ancient crystalline schists, which falls seawards by one or more abrupt cliffs, leaving only a narrow belt of lowland along the coast. Towards the north the lowlands are reduced to a minimum, and the outer plateau escarpment rises almost at once from the sea. The mountain ridges, which run generally parallel to the coast, nowhere reach a great height, the Serra do Crystal or Crystal Range, almost on the equator, which is one of the best marked, being only a little over 4000 feet. At the southern extremity (beyond the latitude of the Congo basin) the highlands broaden out into a plateau (that of Bihe) over 5000 feet high, which is perhaps the most extensive area of a similar altitude in the whole of West Africa. In the south-eastern parts of the Congo basin the ground again rises above the general level of the plateau, and is much more hilly than the greater part of the basin. The Kundelungu plateau, a strip of highland which runs from north to south, west of the head-stream, seems to attain the greatest altitude.

The Congo. The most remote head-stream of the Congo, known as the Chambezi, rises far from the west coast on the high tableland between Lakes Nyasa and Tanganyika, already dealt with in Chapter XI. Descending gradually across nearly level plains in a south-west direction, it reaches a swampy district, which in the rains forms part of the reedy lake of Bangweolo, but dries up in the dry season. On emerging from the swamps under the name Luapula it bends suddenly north, and this direction, or one approaching at times to north-west, it maintains throughout 12° of latitude; a decidedly westward bend being first made almost on the equator. On first turning north the river forms several rapids, but before entering Lake Mwero passes through a swampy tract. The lake itself frequently overflows its low banks in the south-west, but farther north it has higher shores. The course of the river becomes again narrow and much broken by rapids, until after receiving the Lualaba from the south-west, and the Lukuga, the outlet of Tanganyika, from the east, it gains the plains covered by the great equatorial forest. The Lualaba is considered by some the true head-stream of the Congo, as its valley forms a more decided trough in the plateau than the eastern branch, which, however, is both longer and brings down more water. The Lualaba, which flows through a series of shallow lakes, gives its name to the united stream below the junction, a name brought into prominence by the last journeys of Dr. Livingstone, who traced the stream to the Arab mart of Nyangwe, but was forced to leave it without solving the mystery of its termination. Below this place the stream attains an imposing size, being often over a mile wide, and generally deep and unobstructed, until on either side of the equator the series of rapids known as the Stanley Falls occurs, completely preventing navigation. Turning thence by a vast bend west, and finally southwest, the river spreads with ill-defined shores over a huge width in a number of shifting channels, separated by wooded islands and sand-banks. It has here the appearance of forming the last remnant of the old inland sea which must have occupied this region. On approaching the western highlands the river again contracts, and after passing the circular expansion known as Stanley Pool it rushes down towards the sea by a succession of furious rapids, which occupy the narrow winding channel cut in course of ages through the old crystalline rocks which once barred its advance. There are intervals of fairly smooth water during the passage through the highlands, but the stream finally becomes broad and calm only after the Yellala Falls, about 120 miles from the sea.

During this distance there are many side creeks and swampy back-waters on each side, but the water is collected at the actual mouth into a single deep and wide channel. Such is the volume of water poured through this into the ocean that a distinct sub-marine channel is revealed by soundings far out from the coast-line, and the effect of the fresh water on the sea is distinguished two or three hundred miles from the mouth. The vast size of the Congo is explained both by the great extent of its basin and by the fact that the greater part falls within the limits of equatorial rainfall broken by no continued dry season. In this respect it resembles the Amazon, the only other continental river of the first class similarly placed with regard to the equator, and the only river in the world which surpasses it in the volume of water poured forth.

Measured along the windings of the stream the length of the Congo cannot fall short of 3000 miles, a length which is exceeded by the Nile alone in Africa, and by five rivers only in the whole world.

Tributaries of the Congo.—The great width of the Congo basin, which extends from north to south over 20° of latitude, stretching a long distance on either side of the equator, gives room for the formation of an unusual number of large tributaries, several of which exceed in length and volume of water any of the secondary streams of the continent. From the southern limits of the basin to a little beyond 5° S. the whole surface partakes of the northward slope, which causes the upper part of the main stream to take that direction. It is formed of a succession of gentle undulations, which run from south to north, and direct the drainage into a large number of parallel channels, which remain separate for a long distance. After passing 5° S., these streams reach the forest-clad plain traversed by the equator, and the direction is suddenly changed, most of them flowing henceforth westward, and thus uniting into a single channel before joining the Congo.

The most important tributary is the Kassai, which rises

in the south-west near some of the head-streams of the
Zambesi, and, like them, flows eastwards towards the
centre of the continent before turning north. The Lulua
and the Lubilash or Sankuru, east, and the Kwango, west
of the Kassai, are others particularly deserving mention.
Before leaving the higher plateau most of the streams
are broken by rapids and falls, especially between 5° and
6° S., where the principal drop seems to occur, but on
reaching the lower plain they are mostly navigable. Below
the Sankuru, which comes in from the east in $20\frac{1}{2}°$ E.
longitude, the Kassai attains a great width, but is shallow
and broken by islands. After receiving the Kwango
(which has already united with several other northward-
flowing streams) the Kassai soon joins the Congo, but in
spite of its increased volume it becomes comparatively
narrow, and therefore very deep at its mouth, where it is
known as the Kwa. On this account it had not been
suspected, until proved by Wissmann in 1885, that it formed
the outlet for the drainage of so large an extent of country.
Besides the southern streams it receives the Mfini on its
north bank, the main branch coming from the east, but
receiving from the north the drainage of a swampy lake
(the only one of importance in West Africa), which has
been named Leopold II., after the King of the Belgians.
Other tributaries which join the Congo south of its great
bend are the Chuapa and Lulongo, both navigable streams
flowing through the equatorial forest, and the Lomami, a
very swift and winding, but generally deep stream which
rises far to the south, but, like the Upper Congo itself,
maintains its northward flow until the equator is passed.

The principal northern tributary (even longer than the
Kassai) is the Mobangi, the course of which is marked
by the sudden change of direction seen in the Congo itself
and so many of its tributaries. Its upper course is formed
by the Welle, discovered from the north by Schweinfurth
in 1870, the termination of which was for some years one
of the greatest problems of African geography. For nearly
1000 miles the general direction is westerly, parallel to
the central part of the Congo, but on reaching the twentieth
meridian the stream suddenly bends south, like the main

river, which it joins almost on the equator. In this lower part the Mobangi is broad and studded with islands, and at low water is shallow enough to make navigation difficult. Higher up many rapids occur at intervals, some of which

Fig. 32. Rapid on the Aruwimi. (After Stanley.)

bar navigation at high, others at low water, and some are quite impassable. Other important northern tributaries are the Aruwimi, which rises near the Albert Nyanza, and joins the main river a little above its most northern point, and the Sangha, which rises in the north-west corner of the basin, and flows in a little south of the equator. The Aruwimi flows generally parallel with the upper course of the Mobangi, and, like it, is broken by rapids; the Sangha partakes of the direction of the lower Mobangi, and though very winding, is navigable for a long distance. Near its mouth there is a large extent of swampy ground.

The Coast Rivers. Of the minor streams of the

West African coast south of the Gulf of Guinea, the three principal (which in Europe would count as large rivers) all resemble the great rivers of the continent in flowing for a part of their course on the interior plateau before passing through the outer ranges. The Ogowe, just south of the equator, by which the northern part of its basin is intersected, takes its rise on a plateau of moderate elevation, which separates it from the Congo basin. After flowing north and north-west in a richly-wooded valley less than 1000 feet above the sea, it breaks in a westerly direction through the Serra do Crystal, and traverses a fairly level plain of some width before reaching the sea. It forms a large delta thrown out beyond the general line of the coast, the most projecting point of which is known as Cape Lopez (Lopo Gonçalves of the old navigators). It is navigable from the sea only as far as the first mountains. South of the Congo the first stream of any importance is the Kwanza, or Coanza, the upper valley of which lies on the inner side of the plateau of Bihe. The main stream is supposed to rise in a small lake. It flows first north-west, then west during its passage through the hilly country, where it is finally broken by the Livingstone Falls. Below this it is navigable, but is very shallow in the dry season, though the current is very strong during the rains. The bar at the mouth, which shifts from year to year, can only be crossed with great caution. The third river, the Kunene, has its source on the Bihe plateau, and flows due south for a considerable distance before turning west to the sea. In its lower course it traverses an arid region, and loses much of its water by evaporation and possibly by divergent channels, so that in the dry season its mouth is almost choked by sand. Navigation is obstructed at all times by a bar and by several falls not far from the sea. Two considerable streams have been brought to light within quite recent years in the territory of the Cameroons near the head of the Gulf of Guinea. They flow in great part through dense forests and are imperfectly known. The lower course of both is broken by falls, as the high land begins very near the coast. The larger of the two, known as the Sanaga, or Lom, rises apparently in the south of Adamawa.

Climate, Flora, and Fauna.—In many parts of Africa we have seen that in respect of climate and vegetation the main differences depend on the distance from the sea, the coast-lands as a whole differing markedly from the interior. In West Equatorial Africa the zones of climate and vegetation follow each other rather from north to south than in the direction parallel to the coast, the comparatively small elevation of the central Congo basin and its rich covering of alluvial soil bringing it more into agreement with the adjacent coast-lands than with other parts of the interior, while the highlands south of the Congo basin also run at right angles to the coast. Thus the variations of altitude accentuate on the whole the effect of latitude.

The central zone of the region embraces between 4° and 5° of latitude on each side of the equator. This area is marked by a generally high but not excessive temperature and by an absence of great variations, either between night and day or between the seasons. It likewise comes within the most rainy zone of the continent and that in which the dry season is reduced to a minimum. These conditions are just those most favourable to the growth of vegetation, the prevailing character of which is that of a dense forest with a tangled undergrowth, and creepers matting together the trees. It is not yet certain how far the forest is continuous over the surface, as the routes of travellers have mostly kept near the streams, where the growth of plants is always most luxuriant, but in any case forest covers a larger proportion than in any other part of Africa, except the Upper Guinea coast-lands. On either side of this central zone, as the rainfall diminishes and the level of the country rises, the forest becomes thinner, and finally is confined almost entirely to the courses of streams, while the intervening undulations are generally grassy. The floors of the valleys are often occupied by swampy meadows, in which the grass reaches a height of 20 feet. Within the region under consideration this zone is much wider in the south than in the north, owing to the greater extent of the southern slope of the basin. It includes the country on the Lower

Congo itself. In the southern parts of the region, especially near the coast, vegetation becomes still more scanty as the arid region of South-West Africa is approached. At the mouth of the Kunene the country is almost desert, and the small coast-streams north of that river have water only at one time of the year.

Palms are particularly abundant in this region, which

FIG. 33.—STREAM IN THE EQUATORIAL FOREST.
(After Von Götzen.)

is characterised above all (like the Guinea Coast) by the occurrence of the oil palm. It is found southwards as far as the Kwanza River on the coast, and almost reaches the southern limit of the Congo basin inland, though it is absent in the south-west parts of the basin. Several kinds of creepers producing indiarubber grow wild in the forests, and the orchilla lichen (producing two kinds of dye) is said to be particularly abundant. The useful trees include the gum copal, the kola nut, ebony, and other timbers.

The forests of West Africa are unsuited to the life of large wild animals, except the elephant and hippopotamus, and such forms as the giraffe, antelope, rhinoceros, etc., do not appear in any numbers except towards the borders of the region. Buffaloes (of a species different from that of South and East Africa) are, however, met with in some parts. It may be, however, that animal life is not so scarce as it appears, being hidden from view by the dense vegetation. On the other hand, certain forms occur which are not found in the more open regions, such as the two man-like apes, the gorilla and the chimpanzee, and others of the monkey family (*Colobus*, etc.), as well as a small musk deer and one or two species of civet cats. The birds also include species peculiar to the West African region, such as the grey parrot, plantain-eater, and hornbill. The domestic animals of West Africa are likewise few, cattle being rarely met with in the central parts of the Congo basin. Goats and sheep, pigs and poultry, are those chiefly kept.

2. *Inhabitants of West Central Africa*

Apart from the scattered tribes of small stature, considered by some to be the last remnants of an aboriginal population, the inhabitants of this region are almost all **Bantus**. The only tribes which seem to belong rather to the Sudan Negroes are the Monbuttu and Niam-Niam, or Zande, in the extreme north-east of the Congo basin, which have been spoken of at page 118. The kinship of the Fan, who have migrated from the interior, and now dwell near the coast just north of the equator, is very doubtful. Their colour is light brown, and they are said to be morally superior to the Negro. Except quite in the south west, the Bantu tribes are all mainly occupied with agriculture, cultivating principally manioc, earth-nuts, the banana (especially in the forest region), in addition to the usual cereals. It is difficult to group the Bantu tribes into broad subdivisions, since, except in the matter of language, there are numberless variations of type. A

distinction may, however, be drawn between the coast
tribes, which have now been in contact with Europeans
for several centuries, and the peoples of the far interior,
who have lived in isolation until recent years, although
they have in some cases reached a certain degree of
culture; the two groups are separated from one another
by a less thickly peopled tract. None of the coast tribes
have of late years been possessed of much influence. In
the interior the people of Lunda, in the Southern Congo
basin, were for many generations subject to a powerful
chief, bearing the hereditary title Muata Yanvo, but his
power has lately declined. The Kioko, or Chibokwe,
west of Lunda, are a most enterprising race of traders.
Farther still from the coast the principal groups are the
Baluba, east of the Middle Kassai, an intelligent, brave,
and trustworthy race, first made known to the outer world
by Pogge and Wissmann in 1882; the Warua, in the
Upper Congo basin, west of Lake Tanganyika; and the
Manyuema, east of the Upper Congo, near Nyangwe.

In the northern parts a number of tribes still practise
cannibalism. They appear to occupy a zone parallel with
the equator, and towards the east touch on the Monbuttu
and Niam-Niam, likewise cannibals.

Until quite lately the influence of the east-coast Arabs
predominated in the whole region bordering on the Upper
Congo (chiefly east of the river) as far as 8° S. When
Dr. Livingstone in 1869 first reached the Manyuema
country, he found them beginning their work of devasta-
tion, in which the Manyuema themselves finally became
efficient allies. But the Arab power was at last broken
by the forces of the Congo Free State in 1894.

The **tribes of small stature** occupy the most inacces-
sible parts of the forest belt, and bear different names in
different parts. In the southern parts of the Ogowe basin,
where they were first brought to light by Du Chaillu, they
are known as Obongo. At the north-east corner of the
Congo basin another tribe, the Akka, or Tiki-tiki, was first
met with by Schweinfurth. Farther south, on the Aruwimi,
Stanley encountered the Wambutti, and within the great
bend of the Congo, German travellers have discovered

other dwarfish races named Batua, or Watwa. Although these tribes are thought by many to be remnants of a former more widespread population, it is possible that they may merely represent sections of the Bantu who have been driven to the forests by more powerful neighbours, and have degenerated owing to the unfavourable surroundings. They all live by hunting game, which, like the Bushmen of South Africa, to whom some suppose them related, they kill with poisoned arrows.

3. *Portuguese West Africa*

History.—The political connection of the Portuguese with Lower Guinea began almost immediately after the first voyages of their navigators along the coast. Shortly after Diogo Cão's discovery of the Congo (1485) the native chief of the territory of Congo south of the river became a nominal convert to Christianity, and the headquarters of his dominion became, under the name San Salvador, a flourishing town with churches and other buildings. Its splendour, however, lasted but a short time, and before the end of the century St. Paul de Loanda, founded on the coast in 1578, had taken its place as the centre of the Portuguese influence. Other settlements (Benguela, etc.) were soon formed on the coast, but although native traders employed by Portuguese principals penetrated into the far interior, effective occupation has, down to the present day, been confined to a strip along the coast, with a general width of less than 200 miles. The farthest outpost has long been the village of Kassange, in the valley of the Kwango, rather over 300 miles inland.

The Congo has never, like the Zambesi on the east coast, supplied the Portuguese with a road towards the interior, and their authority was first extended over the coast immediately south of the river after Stanley's great journey down it. By recent international agreements the Portuguese sphere has been extended far into the interior, reaching—where widest as far as the Upper

Kassai, and including a large portion of the territory of Lunda. A small isolated territory north of the Congo mouth has also been acquired. Southwards the Lower Kunene forms the boundary with German South-West Africa.

The name Angola, which originally belonged to the territory in the immediate neighbourhood of St. Paul de Loanda, has gradually acquired an extended significance, and is now used officially to include the whole of the Portuguese possessions on this coast. It now includes five provinces, the three older ones of Loanda, Benguela, and Mossamedes (named after the three chief ports) towards the south, and the newer territories of Ambriz and Congo towards the north.

Towns and Settlements.—The Portuguese settlements are almost entirely confined to the coast and to the neighbourhood of the chief trade routes leading from the ports into the interior. The oldest town and the capital of the whole territory is St. Paul de Loanda (15,000), which stands on a small bay sheltered by a long sandy island. The coast here departs for a space from its normal direction and faces north-west, a circumstance which affords additional protection against the south-west winds. Its trade, which in former days consisted chiefly in the export of slaves to Brazil, has increased of late years and exceeds that of any other port in Angola. Two principal routes lead into the interior from this part of the coast. The one proceeds by land, a little north of the Kwanza, to Ambaca (2350 feet above the sea on the second step of the plateau), and Malanje (3800 feet). The other makes use of the Lower Kwanza, on which steamers ply as far as Dondo, at the outer edge of the plateau, and leads thence near the north bank of the Kwanza by Pungo Andongo likewise to Malanje. These routes are continued eastwards to Lunda (*via* Kassanje) and various parts of the Southern Congo basin. A railway has already been opened from Loanda to Ambaca, the inhabitants of which have been noted for their energy as traders. This will lessen the importance of the water route, especially as the Kwanza valley is very unhealthy.

Benguela (properly San Felipe de Benguela), which dates from 1617, has merely an open roadstead, but does a fair amount of trade. It is the terminus of a trade route leading into the interior over the high plateau of Bihe, on which an American mission is at work. The Bihe people are known far and wide as keen traders. Mossamedes, farther south, is of much later date, having been founded only in 1840. It lies on Little Fish Bay, an excellent harbour. The climate is generally healthy, owing to the influence of the cold current which flows up the coast, but the arid nature of the surrounding country is a drawback. A route leads across the Serra da Chella into the Kunene basin, where a colony was founded at Humpata by Boer immigrants from the south in 1881. Many have, however, lately returned southwards. The farthest Portuguese post is at Humbe, near the right bank of the Kunene.

In the north the chief settlement is Ambriz, which was the frontier town of the old province of Angola. Although much of its trade with the interior has been diverted to the Lower Congo, it still exports the produce of the coast-lands in some quantity.

Resources.—Although still but slightly developed, the resources of Angola are considerable. The Kwanza valley especially has much fertile soil suitable for tropical cultivation, and coffee is largely grown in the hilly parts. Cotton is the chief crop in the province of Mossamedes. Rubber (the produce of the forest) and palm oil are largely exported, and some ivory, but this last has lately declined. Copper and other minerals exist, but attempts to work them have met with small success. The volume of trade is many times greater than that of Portuguese East Africa, and the largest part of it is in the hands of British merchants.

4. *The Congo Free State*

History.—As a political unit the Congo Free State is quite without a parallel in history. Instead of being gradually developed from small beginnings through long periods of time, it has at one leap taken its place as a duly organised State with a fixed government and constitution and a territory of nearly a million square miles, the greater part of which had never before 1877 been visited by civilised men. This State is the outcome of the International Association for the opening up of Africa, founded by King Leopold of Belgium, which commenced operations on the west coast in 1879, and which had under Mr. Stanley already accomplished much (without a political status) in the way of establishing stations and opening up a road to the interior when the International Conference met at Berlin in 1884. During the sitting of the Conference, and in some measure as a result of its deliberations, the various nations of Europe, together with the United States, agreed to the creation of an independent State on the Congo under the sovereignty of King Leopold, and its frontiers were laid down so as to include the greater part of the basin of the river as far as Lakes Tanganyika, Mwero, and Bangweolo, and 4° N. latitude. After much negotiation with France and Portugal, the claims of the State to the north bank of the Lower Congo were recognised, and access was thus gained from the sea to the interior. A base of operations had already been formed by the establishment of the station of Boma, about 60 miles from the sea, accessible to large vessels. It is now the capital of the State. Farther inland all the land on the right bank was retained by France.

Communications and Stations.—The chief hope for the development of the country lay in the exceptionally favourable waterways offered by the Congo and its tributaries when once the region of the cataracts is passed, amounting, it is calculated, in all to at least 6000 miles. The first care of the pioneers had therefore been the opening up of a good road to connect the lower reaches of the river (navigable as far as Vivi on the north bank) with Stanley Pool, on which all these navigable channels converge. On the south bank of the Pool a station was founded and named Leopoldville. Steamers were then transported along the road and launched above the cataracts, and by their aid the various streams were

explored and stations founded on their banks, whence an
advance has since been made by land into the more remote
parts. On the main river the station of Bangala, at the
north-west angle of the great bend, is one of the most
flourishing in addition to Leopoldville.

The system of the Kassai is especially serviceable as a
means of communication, as it not only affords a nearly
direct route eastwards towards the region of the Upper
Congo, but by its numerous branches gives access to a
wide area in the southern parts of the State. The station
of Luluaburg, founded near the head of navigation on the
Lulua (an eastern tributary of the Kassai), has been an
important centre, whence many expeditions have set out
towards the Upper Congo region, especially the copper-
bearing region of Katanga, between the Upper Lualaba
and Luapula rivers. Here the kingdom of the tyrannical
ruler Msidi—originally a trader from Unyamwezi in East
Africa—has fallen under the sway of the State. In the
eastern parts of the territory the Arab slave-traders proved
formidable antagonists. The station founded by the
State officials in the neighbourhood of Stanley Falls was
destroyed by them, and a state of war ensued, resulting
at last in the complete overthrow of the Arabs in 1894.
Their trading centres of Nyangwe and Kasango in the
Manyuema country promise to become important centres
of legitimate commerce. The State officials have likewise
pushed up the northern tributary, the Mobangi, and
stations have been founded in the neighbourhood of the
Nile basin. For administrative purposes the State is
divided into sixteen districts, chiefly according to the
navigable waterways which give access to them.

The Congo Railway.—In order to improve the
connection between the Upper and Lower River, a railway
is in course of construction running generally parallel to
the section obstructed by cataracts. The broken nature
of the country has caused many difficulties and progress
has been slow, but about 100 miles out of a total of 260
have been opened for traffic, which has already attained a
considerable volume.

Resources and Trade.—The resources of the Congo

basin have hitherto consisted chiefly in the natural
products of the region, both animal and vegetable. Ivory
is still plentiful, owing to the recent date of the opening
up of the country. The forests abound in creepers producing indiarubber, and the oil palm, gum-copal tree, and
orchilla lichen supply other articles of export. A beginning
has, however, been made with the cultivation of coffee,
which promises to succeed well. The port of entry for

FIG. 34.—SCENERY NEAR THE LOWER CONGO, SHOWING A PART OF THE
CONGO RAILWAY. (Photograph by Captain Weyns.)

the river is Banana, on the north side of its mouth. Trade
is increasing, and will probably attain much larger proportions on the completion of the railway. The greater part
is carried on with Belgium, of which the State is practically
a dependency, although nominally the only connection is
through the person of the ruler.

5. *French Congo*

The nucleus from which the French possessions in
Lower Guinea have spread over a large extent of country
in the Congo basin and adjoining regions to the north

was a small settlement formed in 1842 on the Gabun, a broad estuary running inland just north of the equator. This is not, as its appearance would seem to indicate, the mouth of a large river, but merely an inlet of the sea which receives several small streams flowing from the neighbourhood of the coast. Although a good harbour, it is therefore useless as a means of communication with the interior, and little advance was made until after 1870, when French travellers began to explore the Ogowe River. From 1875 onwards the work was taken up by Savorgnan de Brazza, who in 1878 crossed the water-parting and discovered streams flowing towards the Congo. From that time to the present French activity has never abated, explorers having pushed far into the interior, and, ascending the Sangha and Mobangi rivers, reached the basin of Lake Chad and established communication with the Benue and Niger. By virtue of these journeys, France has established her claim to the whole territory as far as Lake Chad itself—to which she has also extended her North African sphere—in the north, and—north of the Mobangi River—up to the boundary of the Nile basin in the north-east. In thus hastening forward into the more remote regions, much of the intervening country has been left unexplored, and all is still but partially developed.

Stations.—Libreville, on the northern bank of the Gabun, is the capital of the French Congo territory, and the outlet for the trade of its northern parts. On the route inland *via* the Ogowe (which is of little use as a waterway) several stations have been founded, such as Lambarene, Bue, Madiville, and Franceville, while on the Upper Alima, by the valley of which the Congo is reached, is the station of Diele. On the Upper Sangha, and near the north-east frontier, military posts have also been formed. The chief port of entry for the southern districts is Loango, formerly the centre of a native kingdom. A little north of it is the mouth of the Kwilu River, known in its upper course as the Niadi, the valley of which affords a route to Stanley Pool, where the station of Brazzaville has been established. There are several stations on it, the principal being Stephanieville, at the point where the Niadi merges into the Kwilu. When this river was discovered it was thought to compare favourably as a route into the interior with the Lower Congo valley, but the Congo railway even now gives the latter the advantage.

Trade and Resources.—Although undoubtedly possessed of great capabilities, French Congo has not yet developed a large trade, the export of palm oil being at present the most important. The unhealthiness of the coast-lands and the dense forests with which much of the surface is covered are a hindrance to European settlement and agriculture, but the natural resources of the country include, besides palm oil and rubber, valuable dye-woods, gums, and resins, while some coffee, cotton, tobacco, cacao, etc., is grown. The greater part of the trade is carried on by British and German merchants.

6. *The Cameroons*

History.—The Portuguese on their early voyages down the west coast of Africa discovered a broad estuary at the angle formed by the northern and southern limbs of the continent, which they named Rio dos Camerones, or River of Shrimps. Like the Gabun, this is not the mouth of a large river, but merely a tidal estuary which receives several small coast-streams. Until quite recent years this estuary remained almost the only point of contact between Europeans and the surrounding territory, the unknown interior here almost touching the coast even so late as 1884. The name Cameroons—also applied to the lofty volcanic peak just north of the estuary—has been extended to include the whole territory lately placed under German rule, of which the river still forms the chief centre. British traders had long been active here, and a request of the native chiefs to be placed under British protection was under consideration, when in 1884 Germany stepped in and took possession of the country. By agreements with Great Britain and France, the German sphere has been extended inland eastward to 15° E. and north-eastward to Lake Chad and the river Shari. The boundary with

French Congo on the south runs inland from the mouth of the Campo in about $2\frac{1}{4}°$ N.

The chief places on the coast are a group of native villages on the Cameroons estuary, where is the headquarters of the Government; Victoria, on Ambas Bay, a fine natural harbour south of the Cameroons peak (13,000 feet), which was formerly occupied by a British mission, but ceded in 1885; Bimbia, on one of the mouths of the Mungo, which comes from the north, and forms a swampy mangrove-covered delta; and Batanga, towards the south of the territory.

Prospects and Resources.—The chief difficulty in the development of the country consists in the dense forests which clothe the coast-lands, and the absence of a good means of communication with the interior. The rivers are much broken by falls and rapids, and are therefore useless for navigation. The hostility of the native tribes, many of which are warlike and independent and desirous of retaining the trade with the interior in their own hands, has also to be reckoned with. Explorers have, however, overcome these obstacles and penetrated inland, where they have found an open and fertile country on the plateaux after the forest belt has been crossed. Here several stations have been established, the most important being that of Yaunde, between the Sanaga and Nyong rivers. A large part of Adamawa, which is nominally subject to the native ruler of Sokoto (p. 95), and of Baghirmi, subject to Wadai (p. 98), fall within the German sphere, but these are still little touched by German influence, and are naturally approached rather by the Benue River.

Much of the soil is exceedingly fertile, and plantations of cacao, tobacco, and coffee have been successfully started. The chief natural products are palm oil, palm kernels, rubber, and ivory, of which a considerable amount is exported. The larger part of the trade is still carried on by British ships. The great obstacle to prosperity is the unhealthiness of the climate, which makes the country unsuited for the permanent residence of Europeans. Even the Cameroons peak, though much more healthy than the low-lying districts, is by no means exempt from malarial fevers.

CHAPTER XIV

AFRICAN ISLANDS

1. *Madagascar*

Position, Extent, and Surface Features.—Although Madagascar must in very remote times have been united with the African continent, it is now separated by a broad and deep channel, so that it forms a very distinct region in itself, differing remarkably from the mainland in many ways. The narrowest part of the Mozambique channel is almost exactly 250 miles wide, and the depth in the centre amounts in places to no less than 1600 fathoms. This is, however, considerably less than the oceanic depths which are found elsewhere at a similar distance from the shores of the continent. On the east Madagascar sinks below the surface of the ocean with great rapidity, being separated from the Mascarene Islands by depths of 2300 to 2400 fathoms, but towards the north the sea is much shallower, and many small islands rise above its surface.

Madagascar is just 1000 miles long, and is only surpassed in extent by the islands of New Guinea and Borneo. In its general form and structure it has certain resemblances to Africa as a whole. A large part of the surface is formed of old crystalline rocks, and traces of volcanic action are seen in the form of extinct cones and craters. The whole of the centre of the island consists of a broad plateau formed mostly of granite, while the lower plains are composed of newer sedimentary strata. In the north the coasts are much indented, forming spacious bays, but in the south-east the coast-line is very straight. There is here, however, a system of lagoons separated from the sea by narrow banks, which extend along the coast for some hundreds of miles. Between the central plateau and the west coast, ranges of hills formed of sedimentary rocks occur, running north and south, and enclosing a longitudinal depression, which extends from the north-west coast southwards for more than half the whole length of the island. Many ridges and peaks rise above the general surface of the plateau, following on the whole a north to south arrangement. The greatest elevations, which rise to a height of 9000 to 10,000 feet, lie near the eastern edge, and the rivers which flow westwards have therefore the longer courses. The four largest are the Ikopa, flowing from the centre of the island north and north-west into Mojanga Bay, and the Tsiribihina, Mangoka, and Ongulahi, flowing generally west in the central and southern parts.

Climate, Flora, and Fauna.—Being surrounded by the ocean, the climate is generally moist and equable. Except in the extreme south, where the precipitation is scanty, most rain falls in the summer months. At this time especially the climate is unhealthy except on the higher plateaux. The vegetation varies to some extent according to the altitude, the granite plateau being generally grassy, or at most dotted over with scattered trees, while the lowlands are clothed with a varied and luxuriant growth. The densest forest occupies a narrow band running quite round the island at the foot of the plateau escarpments, but touching the coast in the north-east only. Both as regards its flora and fauna, Madagascar is remarkable for its peculiar forms, which do not occur anywhere else on the globe, a fact which shows that the separation from the continent must have taken place in very remote times. The species of plants are very varied. One of the characteristic trees is the Ravenala, or Traveller's Tree, so called from the supply of water to be obtained from the hollow leaf stalks, which are arranged in a fan shape at the top of the stem. It is allied to the banana. Other plants are allied to African, Asiatic, or even South American types. All the larger wild animals of Africa are entirely wanting, and it seems likely that the forms which occur are those which peopled Africa and the Old World generally before the existing African mammals were introduced. This will explain the fact that many of the forms have allies in Asia and the Malay islands as well as in Africa. Thus the lemurs, a group of four-handed animals allied to the monkeys, though not found in such numbers in any other part of the world, range from West Africa to South-East Asia. They live in the dense forests, hardly ever leaving the trees, and roaming about chiefly by night. Other animals belong to the groups of the insect-eaters, civet cats, and rodents, and there is also a peculiar cat-like animal named *Cryptoprocta*. None of these animals are found in other parts of the globe.

Inhabitants.—The native races of Madagascar likewise differ from those of the neighbouring continent, for it is only on the west coast that any trace of the Negro race is found. Elsewhere the people seem rather related to the Malay family, though separated from its other representatives by the whole breadth of the Indian Ocean. The ocean current which sets across this from east to west has probably assisted the migration of members of this family towards Madagascar in remote times. The tribes most decidedly Malayan in type live in the centre and the east of the island, the Hovas in the centre being the most important. These have, during the nineteenth century, imposed their authority on the other inhabitants, including the Sakalavas on the west coast, who show a large admixture of Negro blood. Other important tribes are the Betsileo, south of the Hovas, and the Tanala, who dwell in the forests towards the south-east. These are less decidedly Malayan than the Hovas, who have shown themselves the most intelligent, and ready to adopt civilised habits, of all the natives of the island. Arab influence is to be traced in the customs and beliefs of the people.

Political Relations.—The Portuguese discovered Madagascar (which had been known to the early Greek geographers as Menuthias) in 1506, calling it the Isle of São Lourenço, but did not enter into permanent relations with the island. The Dutch, and after them the French, established posts on the east coast, and French agents even explored the interior in the eighteenth century, but without permanently establishing French influence, except at the small islands of Ste. Marie and Nossi-bé. After the Hovas had established their power early in the nineteenth century, British missionaries were encouraged to work in the country, and in 1869 Queen Ranavalona II. formally embraced Christianity, which is now professed by over a quarter of a million of the people. The greater part belong to the English Protestant sects, but some are converts of a Roman Catholic mission. In 1883 the French began a new period of activity by making war on the Malagasy, but without much result beyond securing a footing in the north and east of the island. In 1890 Great Britain recognised a French protectorate over Madagascar, which was imposed on the Hovas by a military campaign in 1895. France has now taken the whole government into her own hands, but discontent is rife, and her hold on the country is only maintained by military force.

Towns, Resources.—The capital of the island is Antananarivo, in the central province of Imerina, the cradle of the Hova tribe. Under the rule of the Hovas its population has reached a total of 100,000, with many fine buildings, including, besides the royal palaces, cathedrals, colleges, hospitals, and even an observatory. It was occupied by the French army in 1895. The chief ports are Tamatave, in the east, whence the route to the capital leaves the coast, and Mojanga, in the north-west. Diego Suarez Bay, in the extreme north, forms a splendid land locked harbour. Fianarantsoa, in the Betsileo country, is the most important native centre in the southern half of the island.

The coast-lands are fertile but unhealthy, whereas the more healthy plateaux produce little but grass, and are often dry and barren. Thus there is little prospect of European colonisation. Rice is the staple crop of the island, and some is exported. Other exports are cattle and hides, rubber, gum copal, hemp, sugar, and coffee. The forests also produce some valuable woods, including ebony. Cotton and other manufactured goods are im ported. The trade with Great Britain has grown rapidly since 1884, and exceeds that with France, but the largest amount is carried on with the islands of Mauritius and Réunion.

The Comoro Islands, between the north end of Madagascar and the continent, also belong to France. The native inhabitants are of the Bantu race. The most fertile is Mayotte, where the sugar-cane is grown to a considerable extent. It has several good anchorages.

2. *Mauritius, Réunion, Seychelles, etc.*

East of Madagascar, but separated from it by a channel over 2000 fathoms deep, there is an extensive submarine ridge or bank covered by shallower water, the highest points of which rise above the surface as islands. This bank runs generally parallel with Madagascar, but towards the north it bends westwards in the direction of the groups north of that island.

The largest islands, Réunion and Mauritius, which are of volcanic origin, lie at the south end of the ridge, and are sometimes known as the Mascarenhas, after the Portuguese captain who discovered them in 1513. They were then uninhabited, but were afterwards settled by the French (Mauritius having been previously occupied for a time by the Dutch, who gave it its present name). In the Napoleonic wars they were seized by Great Britain, but Réunion (previously known as Bourbon) was restored to France at the peace of 1814.

Mauritius.—Although it is now a British colony, the bulk of the white residents in Mauritius are still of French extraction, and the French language is spoken. The highest point of the island is about 2700 feet above the sea. It has a plentiful rainfall, and the mountainous parts are thickly wooded. The lower lands have been much cleared for the planting of sugar, which is the principal crop. Hindu coolies have been introduced in large numbers as labourers, and there are also some

Chinese as well as Africans. Great damage is often done by cyclones, to which the island is very liable. The capital of Mauritius is Port Louis (58,000), on the north-west coast, whence railways cross the island to Port Savanna and Mahebourg, on the south-east, and Grande Rivière, on the east. The annual value of the exports, which go chiefly to Great Britain, South Africa, Australia, and India, has in some years exceeded £2,000,000.

Dependencies of Mauritius are the small island of **Rodriguez**, lying at a considerable distance to the east, and several groups of small islands on the bank already spoken of. The principal is that of the **Seychelles**, north-east of Madagascar, the largest of which is Mahé, containing Port Victoria. Coco-nut oil, copra (the dried fruit of the coco-nut), vanilla, etc., are exported.

Réunion.—The French island of Réunion is very similar to Mauritius in its general character, but is more mountainous, and in the Piton des Neiges reaches a height of 10,000 feet. Sugar is here too the most important product.

The capital is St. Denis, on the north coast, and St. Pierre, on the south coast, is the next town in importance. A railway starting thence goes nearly all round the island.

3. *West African Groups*

Off the north-west coast of Africa, and farther out in the North Atlantic Ocean, there are several groups of islands which played an important part in the early voyages of discovery, as they afforded ports of shelter and means of obtaining water, which was not to be found on the arid coast opposite. One of these groups—the Canaries, the nearest to the African coast—was known to the ancients as the Fortunate Islands, but even these remained almost an unknown land in Europe until visited by the voyagers of the fourteenth and fifteenth centuries. All the groups are principally of volcanic origin.

The **Canaries** consist of seven principal islands, just south of 30° N. The five largest form a curve open to the north, the easternmost being about 60 miles from the coast of Africa. The best-known are Teneriffe and Gran Canaria, farther west, which, like the others in that direction, rise abruptly from the water into lofty peaks. The famous peak of Teneriffe is a tapering volcanic cone over 12,000 feet high. The climate of the western islands is healthy and fairly temperate. The rainfall is generally scanty, especially in the east of the group and on the lower grounds, where the most characteristic plants are Euphorbias and the Dragon tree, both African in type. On the higher grounds European trees flourish, such as the laurel, pine, juniper, oak, and chestnut. The fruit trees of South Europe grow well. The principal domestic animal is the goat. The aboriginal inhabitants were the Guanches, supposed to be allied to the Berbers. At the present day there is a very mixed population, including representatives of various European nations, besides the Spaniards, who have held the islands since the end of the fifteenth century. The principal ports are Santa Cruz, on Teneriffe, the seat of Government and an important coaling station, and Las Palmas, on Gran Canaria.

Madeira, with the smaller island of Porto Santo, lies farther from the African coast, and was on this account first discovered at a later date than the Canaries. During the life of Prince Henry of Portugal it was colonised by the Portuguese, and has ever since belonged to Portugal, of which it is considered to form a part. The island is traversed throughout its length (from east to west) by a mountainous ridge rising above 6000 feet. It has a very temperate climate and is much visited by invalids from Europe. It produces fruit in abundance and is famous for its wines. The capital is Funchal, on the south coast.

The Azores lie so far out in the Atlantic Ocean that they have little in common with Africa. Like Madeira, they form an electoral district of Portugal, and their connection with that country likewise dates from the time of Prince Henry. They form a scattered group of eight principal islands, surrounded on all sides by the ocean depths. They are mostly rugged and mountainous—the highest point of Pico being 7600 feet above the sea—and were originally clothed with forest. They are well watered and fertile, and produce various fruits, wines, tobacco, etc., which form the principal exports. The fauna is very limited, except in the case of the birds, whose power of flight has enabled them to reach this remote group. The capital is Angra, on the island of Terceira, and other towns are Ponta Delgada and Ribeira Grande, both in São Miguel (St. Michael), the largest island. The inhabitants are Portuguese with a mixture of Negro blood.

The Cape Verde group is composed of a greater number of islands than either the Azores or Canaries, but many of them are very small. They are arranged in the form of an irregular horseshoe, of which the northern branch bears the name of the Windward, the southern that of the Leeward islands. The most mountainous are Santo Antão, the most westerly of the Windwards, and Fogo and Santiago, in the centre of the Leewards. Fogo has an active volcano rising to a height of 8800 feet. Both the flora and fauna are scanty. The group falls within the zone of tropical rains, limited to the late summer and autumn, the rest of the year being dry, a fact which causes them to suffer severely from drought at times. In the rains the climate is unhealthy. Some tropical and South European crops are grown, and chinchona plantations have lately been formed, but the exports are not large. The principal importance of the group consists in its position on the ocean routes from Europe to South Africa and Brazil. It was usually touched at by the old voyagers *en route* for these countries, and at the present day Porto Grande, on the barren island of São Vicente, is an important coaling station with an excellent harbour. The group belongs to Portugal, of which it forms a colonial Government. The capital is Villa da Praia, on Santiago.

4. *Islands in the South Atlantic*

In the South Atlantic Ocean a few small volcanic islands rise far away from land, from the ocean depths which surround them on all sides. Two of these, Ascension and St. Helena, are quite isolated submarine peaks, while farther south the Tristan da Cunha group consists of but three islets, separated by 280 miles of sea from the solitary Gough Island, to the south-east. Ascension and Tristan da Cunha rise from the submarine ridge which divides the South Atlantic into an eastern and a western basin, while St. Helena lies a little off it to the east.

Ascension, which is about 8 miles long, rises in the south-east to a height of 2800 feet. The surface is rugged and nearly devoid of vegetation, as water is extremely scarce. Goats and poultry have been introduced, and the former have become wild. Turtles visit the island in great numbers to lay their eggs, and are caught for the sake of their flesh. Ascension was occupied by Great Britain in 1815, and has a population of about 200, who supply provisions to passing ships. It serves as a naval station for the West African squadron.

St. Helena measures about 10 miles by 8, and is likewise very rugged and bounded by vast precipices. The highest point is about 2700 feet above the sea. It has a plentiful rainfall, and was formerly clothed in luxuriant vegetation, but the destruction wrought by goats and by the reckless cutting of the timber has reduced it to a state of barrenness. Being no longer protected by vegetation, the soil is washed away by the rain, so that efforts to introduce new plants have met with little success. The chief crop is potatoes. A vigorous effort is now being made to develop the fisheries. St. Helena was discovered by the Portuguese in 1502, but came into the hands of the East India Company in 1651, and is now a Crown colony. Its importance has very much declined since the opening of the Suez Canal, which now diverts a great part of the shipping that formerly touched at St. Helena on the voyage to the East. Jamestown, on the north-west coast, is the capital, and serves as a coaling station for the British Navy.

Tristan da Cunha rises in a regular cone to a height of 8300 feet. The climate is mild and damp, and a limited number of plants grow on the island, which supports some cattle, sheep, and poultry. Potatoes are here too the principal crop, but some maize and European fruits are also grown. The population is about 100, composed of settlers of various nationalities, who consider themselves as under the protection of Great Britain.

5. *Islands in the Gulf of Guinea*

Near the head of the Gulf of Guinea four small islands lie in a straight line from south-west to north-east, which is continued on the mainland by the Cameroons Mountain. All are of volcanic origin, and evidently belong to the same line of volcanic disturbance.

Fernando Po, the largest of the four, and the nearest to the mainland, is clothed with rich vegetation, and rises in Clarence Peak to a height of over 10,000 feet. The native inhabitants are known as the "Bubi." It was discovered by the Portuguese, but was ceded to Spain in 1778. The only town is Santa Isabel, on the north coast, formerly called Clarence Town, which was for a time occupied by Great Britain. The soil is fertile, and tropical products are successfully cultivated, but the climate is not healthy.

Prince's Island and St. Thomas, which come next in order, are still retained by Portugal. Both are extremely fertile, but St. Thomas is the larger and more important. It rises on the west side to a height of 7000 feet. Its climate is comparatively healthy, and the Portuguese settlers have thriven, and own flourishing plantations of sugar cane, coffee, and cacao, while chinchona trees have lately been introduced, and have done well. The capital is Cidade, on the north-east coast.

Annobon, the last of the four, was ceded to Spain with Fernando Po. It is very small, and is inhabited only by the descendants of Negro slaves. The principal produce consists of oranges and other fruits.

SUMMARY OF THE GEOGRAPHY OF AFRICA

A. Physical Geography

Position and Extent (p. 1)

AFRICA forms the south-western section of the land-mass of the Old World, being separated from Europe, on the north, by the Mediterranean Sea; and from Asia, on the north-east, by the Red Sea. Elsewhere it falls rapidly to great ocean depths (Indian Ocean on the east, Atlantic on the west).

Area, 11,900,000 square miles, or greater than that of any other continent except Asia. *Length*, north to south, 5000 miles; east to west, 4600 miles.

Outline, etc. (p. 2)

Outline very uniform. Two main segments at right angles. Few gulfs or peninsulas.

Seas and Gulfs—
  North: Mediterranean Sea; Gulfs of Gabes and Sidra (the "Syrtes").
  East: *Red Sea:* Gulf of Suez; *Indian Ocean:* Gulf of Aden, Mozambique Channel (between Africa and Madagascar).
  West: *Atlantic Ocean:* Gulf of Guinea, with Bights of Benin and Biafra.

Bays—
  North: Gulf of Bougie, Gulf of Tunis, Gulf of Hammamet.
  East: Tajura Bay (in Gulf of Aden), Formosa Bay.
  South-east: Bay of Sofala, Delagoa Bay.
  South: Algoa Bay, False Bay.
  South-west: Table Bay, Saldania Bay, St. Helena Bay, Walfish Bay.
  West: Little Fish Bay, Bengo Bay, Lopez Bay, Corisco Bay, Yawri Bay, Arguin Bay, Rio del Oro.
  In Madagascar: Antongil Bay, St. Augustin Bay, Mojanga Bay.

Peninsula.—The "Eastern Horn."

Capes—
  North : Cape Tres Forcas (in Marocco), Capes Blanc and Bon (in Tunis).
  East : Ras Benas and Ras Ranai (in Red Sea) ; Cape Guardafui, Ras Hafun (on "Eastern Horn") ; Cape Delgado (near 10° S.) ; Cape Corrientes (near Tropic of Capricorn).
  South : Cape Agulhas, Cape of Good Hope.
  West : Cape Frio (near 19° S.) ; Cape Lopez (near Equator) ; Cape Formoso (Niger delta) ; Cape Three Points, Cape Palmas (Guinea Coast) ; Cape Verde, Cape Blanco, Cape Bojador, Cape Juby, Cape Nun, Cape Spartel (North-West Coast).

Islands—
  East Coast : Sokotra, Pemba, Zanzibar.
  Indian Ocean : Madagascar, Comoro Islands, Réunion, Mauritius, Rodriguez, Seychelles, Amirante Islands, Aldabra.
  South Atlantic : Tristan da Cunha, St. Helena, Ascension.
  Gulf of Guinea : Fernando Po, Prince's Island, San Thomé, Annobon.
  North-West Coast and North Atlantic : Cape Verde Islands, Canaries, Madeira, Azores.

Structure and Relief (pp. 2, 3)

Abundance of older rocks as compared with more recent formations.

General absence of low plains. High interior plateau, bounded by steep escarpments not far from the coast. Lower northern and higher southern section. Southern section broken in west by hollow of Congo basin.

Mountain Ranges few. Isolated groups more frequent.
  North : Atlas (Jebel Aiashi, 14,000 feet) ; Air and Tibesti groups, in centre of northern plateau.
  East : Mountains of Abyssinia (Ras Dajan, Abba Yared, 15,000 feet) ; East African volcanic peaks —Kilima Njaro, Kenya, Ruwenzori (18,000 to 19,000 feet), Elgon, Kirunga ; groups in interior of Mozambique—Namuli Hills, Mlanji, Zomba (8000 to 10,000 feet).
  South-east and South : Matoppo Hills, Drakensberg, Nieuwveld.

West: Serra do Crystal (near Equator), Cameroons (at head of Gulf of Guinea), Futa Jallon Highlands (near western extremity).

In Islands: Peak of Teneriffe (12,300 feet), in the Canaries; Mountains of Madagascar (Tsiafajavona, 8600 feet); Piton des Neiges (10,400 feet), in Réunion.

Rivers and Lakes (p. 7)

The principal *rivers* flow for long distances across the interior plateau; the secondary streams mostly rise in its outer escarpments.

1. Flowing into the Mediterranean—
   Nile (4000 miles), flowing across interior plateau (p. 107).
   Muluya, in Marocco ⎫
   Sheliff, in Algeria ⎬ Coast streams.
   Mejerda, in Tunis ⎭
2. Flowing into the Indian Ocean—
   Zambesi (2000 miles) (p. 164) ⎫
   Limpopo (900 miles) (p. 185) ⎭ Chiefly on interior plateau.
   Jub ⎫
   Tana ⎪
   Rufiji ⎬ Coast streams.
   Rovuma ⎪
   Sabi ⎭
3. Flowing into the Atlantic Ocean—
   Congo (3100 miles) (p. 208) ⎫
   Niger (2600 miles) (p. 88) ⎬ Chiefly on interior plateau.
   Orange (1200 miles) (p. 184) ⎭
   Kunene ⎫
   Kwanza ⎪
   Ogowe ⎬ In southern section
   Sanaga or Lom ⎭
   Volta ⎫
   Akba or Komoe ⎬ On Guinea Coast ⎬ Coast streams.
   Bandama ⎭
   Gambia (p. 87)
   Senegal (900 miles) (p. 87)
4. Not reaching the sea—
   Shari, flowing into Lake Chad (p. 89).
   Webi Shabeli, ending in a swamp close to the East Coast.
   Tioghe or Kubango, flowing into Lake Ngami.

Lakes—
1. In centre of northern plateau—
    Chad (p. 89), without permanent outlet.
2. On high eastern plateau—
    Tana, in Abyssinia
    Victoria Nyanza (p. 108)
    Albert Nyanza (p. 108)   } Nile system.
    Albert Edward Nyanza (p. 108)
    Rudolf (Basso Norok), without outlet.
    Tanganyika (p. 147)
    Mwero (p. 209)   } Congo system.
    Bangweolo (p. 209)
    Nyasa (pp. 147, 168), Zambesi system.
    Shirwa, without outlet.
3. In West Equatorial Africa—
    Lake Leopold II., Congo system.
4. On South African plateau—
    Ngami, without permanent outlet.

Climatic Regions (p. 10)

Rainfall generally scanty in north, more abundant in centre and south.
1. Northern temperate region (pp. 62, 64). Rain in winter, spring, or early summer. (Similar to South Europe.)
2. Northern deserts (p. 76). Almost rainless; great variations in temperature between the seasons (intensely hot in summer), and between day and night.
3. Tropical zone (the greater part of Africa between 15° N. and 20° S.). Year divided into dry and rainy seasons.
4. Equatorial zone. Most pronounced in West Africa (pp. 92, 214). A large amount of rain throughout most of the year. Small variations in temperature.
5. Arid region in South-West Africa (p. 186).
6. South temperate zone (p. 186). Rain in winter.

Vegetation (p. 11)

1. Northern Africa. Plants of South Europe.
2. Northern deserts (Sahara). Little vegetation, highly specialised.
3. Region of savannahs, interspersed with trees, occupying most of tropical Africa. Baobab found everywhere.

SUMMARY

3*a*. Steppe region of East Africa (Somaliland, etc.). Thorny jungles.
4. Forest region of West African coasts and Congo basin. Oil palm.
5. Arid region of South-West Africa.
6. South temperate region. Heaths.

Animals (p. 17)

1. North of Tropic of Cancer. Mixture of European and Southern forms.
2. Northern desert region. Camel.
3. East and South Africa. Abundance of large animals: lion, leopard, antelope, giraffe, zebra, elephant, rhinoceros, hippopotamus, crocodile, ostrich, etc.
4. West African forest region. Few large animals except the elephant, hippopotamus, and crocodile. Man-like apes.

Productions

1. *Animal Productions—*
   Ivory: the central parts of tropical Africa.
   Ostrich feathers: South Africa, southern borders of Sahara.
   Wool (merino sheep and Angora goat): South Africa.
   Hides: North and South Africa.
   Wax: Madagascar, Gambia, West Central Africa.
2. *Vegetable Productions—*
   Wheat: Egypt, Algeria.
   Maize: Egypt, South Africa, parts of the tropical zone, especially in South-East Africa.
   Rice: Madagascar, and other parts of the tropical zone.
   Millet (pennisetum): tropical zone generally.
   Durra or Guinea corn (sorghum): especially in the Central Sudan.
   Banana or plantain: the hot and moist regions.
   Dates: the Saharan oases.
   Coffee: Galla countries (south of Red Sea), region south of Lake Nyasa, parts of West Africa (Angola, Liberia, etc.).
   Sugar: Natal, Egypt, tropical coast-lands.
   Wine: South Africa, Algeria.
   Cotton: Egypt, tropical zone.
   Palm oil: West Africa (Guinea Coasts and Congo basin).

Rubber: West African forest region, South-East Africa.
Ground nuts (*arachis*): tropical West Africa, especially Senegal.
Gums: the Eastern Sudan, Somaliland, Senegal.
Alfa grass: North Africa (Algeria, Tripoli).
Indigo: tropical Africa generally.
Orchilla lichen: tropical Africa, especially Congo basin.
Cochineal: Canary Islands.

3. *Mineral Productions—*
Gold: South and South-East Africa, Gold Coast, Galla countries.
Copper: South Africa, Algeria, Marocco, South-East Congo basin, Darfur.
Iron: Marocco, Algeria, Orange Free State, and (in small quantities) throughout most of tropical Africa.
Coal: Cape Colony, Natal, Orange Free State.
Diamonds: South Africa.
Salt: the Western Sahara.

Inhabitants

I. *The Northern Races* (p. 20)—
  1. Semites:
    Arabs, in North Africa (Tripoli, etc.), the Western Sahara, the Upper Nile.
    Abyssinians.
  2. Hamites:
    Berbers, in Marocco and Algeria.
    Fellahin of Egypt.
    Tuareg, in Central Sahara.
    Bejas, between the Nile and the Red Sea.
    Agaos, in Abyssinia.
    Afar or Danákil, between Abyssinia and the Red Sea.
    Somalis and Gallas, in the Eastern Horn and regions south and west.

II. *Mixed and Doubtful Races* (pp. 93, 216), probably mostly a mixture of Hamites and Negroes—
  Tibu or Teda, in Central Sahara.
  Nubians, on the Nile.
  Fulahs or Fellata, in the Western Sudan.
  Masai, in East Africa.
  Fan, on West Coast near the Equator.

SUMMARY

III. *The Southern Races* (p. 21)—
  1. Sudan Negroes.
    (*a*) With a slight admixture of Hamite blood—
      Wolofs, between the Senegal and Gambia.
      Mandingos of the Western Sudan.
      Hausas of the Central Sudan.
      Kanuri of Bornu.
    (*b*) Pure Negroes—
      "Chi"-speaking peoples (in Ashanti).
      "Ewe"-speaking peoples (in Dahome).
      Yorubas, west of Lower Niger.
      Nilotic branch, on Upper Nile and Western affluents (Shilluk, Dinka, Bongo, etc.).
    (*c*) More nearly allied to the Bantus—
      A-Zande or Niam-Niam, and Monbuttu, on Nile-Congo watershed.
  2. Bantu Negroes—
    Bateke, Balolo, Baluba, Waregga, Manyuema, Warua, etc., Congo basin.
    Amboellas, Barotse, Mambunda, etc., Upper Kubango and Zambesi.
    Waganda, Wanyamwezi, etc., East Africa.
    Bechuanas and Basutos, South Central Africa.
    Zulus, Kaffirs, Matabili, Angoni, etc., South-East Africa.
  3. Negritoes (affinities doubtful)—
    Hottentots } South-West Africa.
    Bushmen   }
    Obongo, Ogowe basin.
    Akka or Tiki-tiki } North-East Congo basin.
    Wambutti   }
    Watwa, Central Congo basin.

B. Political Divisions

BRITISH POSSESSIONS

1. *In West Africa* (Sudan) (p. 98)—
    Gambia (4000 square miles).
      Chief Town: Bathurst (6000).
      Products: ground nuts, bees-wax, rubber.

Sierra Leone (27,000 square miles).
Chief Town: Freetown (30,000).
Products: palm oil and kernels, kola nuts, rubber, gum-copal, etc.

Gold Coast (50,000 square miles).
Towns: Cape Coast Castle (10,000), Accra (20,000), Axim, Elmina, Addah, Quittah.
Products: palm oil and kernels, rubber, gold dust, timber.

Lagos and Yoruba (25,000 square miles).
Chief Towns: Lagos (75,000), on coast; Ibadan (150,000), Oyo (70,000), in interior.
Products: Palm oil and kernels.

Niger Coast Protectorate (50,000 square miles).
Chief Towns: Bonny, Duke Town, Creek Town, Okrika.
Products: palm oil and kernels, rubber, ebony, camwood, etc.

Niger Company's Territories (375,000 square miles).
Chief Towns:—
  (*a*) In territory administered by the Company—Asaba, the capital; Akassa (at mouth of Niger), Lokoja, Egga.
  (*b*) In Empire of Sokoto—Wurnu, the capital; Sokoto, Kano (60,000), Katsena, Zaria, Yakoba, Gombe, Muri, Keffi, Yola. In Bornu—Kuka.
Products: gums, hides, rubber, ivory, palm oil and kernels, vegetable butter, etc.

2. *In North-East Africa* (p. 141)—
British Somaliland (67,000 square miles).
Chief Towns: Berbera (30,000 at times), Zaila, Bulhar.
Products: cattle and sheep, hides, gums, ostrich feathers.

Sokotra.
Chief Town: Tamarida.
Products: cattle, Socotrine aloes.

3. *In East Africa* (p. 155)
Zanzibar and Pemba Islands (1000 square miles).
Chief Town: Zanzibar (100,000).
Products: cloves, copra; (the exports also include mainland products, ivory, rubber, etc.).

Mainland Territory (Ibea) (150,000 square miles).
Chief Towns: Mombasa (20,000), Wanga, Freretown,

Malindi, Lamu, Kismayu. Stations in interior: Machako's, Kikuyu, Eldoma, Ports Victoria and Alice, Fort Kampala (at Mengo, capital of Uganda).
Products: Sesame seed, ivory, copra, rubber, orchilla lichen, gums, hides, etc.

4. *In the Zambesi Basin* (p. 174)—
   British Central Africa (protectorate and territory under the South Africa Co.) (280,000 square miles).
   Settlements: Blantyre (6000), Zomba, Chiromo, Fort Johnston, Kota-kota, Bandawe, Karonga, etc. (in the Protectorate); Abercorn, Rhodesia, Fort Rosebery, etc. (in the South Africa Co.'s territory).
   Products: Coffee, rice, wheat, sugar, cotton, rubber.
   Matabili and Mashona Lands (100,000 square miles).
   Towns: Bulawayo, Salisbury, Tuli, Victoria, Charter.
   Products: Gold and other minerals.

5. *In South Africa* (pp. 191, 200)—
   Colony of the Cape of Good Hope (226,000 square miles ; 1½ million inhabitants).
   Chief Towns: (a) Ports—Cape Town (51,000), Port Elizabeth (23,000), East London, Port Alfred ; (b) Interior Towns—Grahamstown (woollen manufactures) (10,500), Kimberley (29,000), Beaconsfield (10,500), Paarl (7700), King William's Town, Graaf Reinet, Worcester, Uitenhage. (In Bechuanaland), Vryburg, Mafeking.
   Products: Wool and hair, ostrich feathers, hides, butter, wheat, maize, kaffir corn (sorghum), fruit, wine, diamonds, copper, coal.
   Natal and Zululand (30,000 square miles) : Natal, 550,000 inhabitants.
   Towns: Durban (28,000), Pietermaritzburg (18,000), Ladysmith, Newcastle, Colenso, Greytown, Weenen, Richmond.
   Products: Wool, hides, sugar, tea, coffee, tobacco, arrowroot, cereals, coal.
   Bechuanaland (118,000 square miles).
   Settlements: Molopolole, Shoshong, Palapye.

6. *Islands* (pp. 231, 234)—
   Mauritius, Rodriguez, etc. (1000 square miles).
   Towns: (in Mauritius), Port Louis (58,000), Port

Savanna, Mahebourg, Grande Rivière, Curepipe. (In the Seychelles), Port Victoria.
Ascension, St. Helena, Tristan da Cunha (130 square miles).
Town: (In St. Helena), Jamestown.
Products: Goats, potatoes.

FRENCH POSSESSIONS

1. *In North Africa* (pp. 69, 85)—
    Algeria (300,000 square miles; 4,000,000 inhabitants).
    Towns: Algiers (82,000), Oran (74,000), Constantine (41,000), Mostaganem, Bona, Philippeville, Bougie, Tlemcen, Sidi-bel-Abbes, Blida, Guelma, Géryville, Batna, Biskra.
    Products: Wheat, wine, garden produce, alfa, copper, iron.
    Tunis (50,000 square miles; 1½ million inhabitants).
    Towns: Tunis (145,000), Bizerta (8000), Susa, Sfax, Kairwan.
    French Sahara (1,600,000 square miles).
    Products: Dates, salt.
2. *In West Africa* (pp. 103, 223)—
    Senegal, French Sudan, French Guinea, Ivory Coast (? 500,000 square miles).
    Towns and Stations: St. Louis (20,000), Dakar, Rufisque, Bakel, Kayes, Bamaku, Segu, Sansandig, Kolodugu, Jenne, Timbuktu, Bandiagara, Konakri, Boke, Victoria, Melakori, Grand Bassam.
    Products: Ground nuts, cereals, rubber, timber, gums.
    Benin Coast with **Dahome** (14,000 square miles).
    Towns: Whyda, Kotonu, Abome.
    Products: Palm oil and kernels, maize, rubber, ivory.
    French Congo, etc. (560,000 square miles).
    Towns and Stations: Libreville, Franceville, Loango.
    Products: Palm oil, rubber, gums and resins, dye-woods, cotton, tobacco, cacao, coffee.
3. *On the Red Sea* (p. 142)—
    French Somali Coast (8600 square miles).
    Towns: Jibuti, Obok.

4. *Islands* (pp. 227, 231)—
   Madagascar (228,000 square miles).
      Towns: (on coasts), Tamatave, Mojanga ; (in interior), Antananarivo.
      Products: Rice, cattle and hides, rubber, gum-copal, hemp, sugar, coffee, ebony, etc.
   Comoro Islands (760 square miles).
      Product: Sugar.
   Réunion (760 square miles).
      Towns: St. Denis, St. Pierre.
      Products: Sugar, coffee, cacao, vanilla, etc.

PORTUGUESE POSSESSIONS

Portuguese East Africa (p. 172) (300,000 square miles).
   Towns: Mozambique, Lourenço Marques, Quillimane, Beira.
   Products: Oil nuts and seeds, rubber, ivory, sugar.
Angola (p. 218) (500,000 square miles).
   Towns: St. Paul de Loanda (15,000), Benguela, Mossamedes, Ambriz, Dondo, Malanje.
   Products: Coffee, cotton, rubber, palm oil, ivory.
Portuguese Guinea (p. 105) (14,000 square miles).
   Town: Bissao.
Cape Verde Islands (p. 234) (1500 square miles).
   Towns: Villa da Praia, Porto Grande.
Madeira (p. 233) (330 square miles).
   Town: Funchal.
   Products: Fruit, wine.
San Thomé and Principe (p. 236) (420 square miles).
   Town: Cidade (on San Thomé).
   Products: Coffee, cacao, sugar, chinchona.

GERMAN POSSESSIONS

German East Africa (p. 160) (350,000 square miles).
   Towns: Bagamoyo, Dar-es-Salaam, Tanga, Kilwa.
   Products: Ivory, rubber, gum-copal, orchilla weed, tobacco, cotton.

German South-West Africa (p. 205) (320,000 square miles).
  Stations: Swakopmund, Sandwich Harbour, Angra Pequeña, Otjimbingue.
  Products: Cattle, copper.

Cameroons (p. 225) (190,000 square miles).
  Towns and Stations: Cameroons, Batanga.
  Products: Palm oil and kernels, rubber, ivory, cacao.

Togoland (p. 106) (20,000 square miles).
  Towns and Stations: Togo, Bagida, Little Popo, Misahöhe, Bismarkburg.
  Products: Palm oil and kernels, ivory, copra.

ITALIAN POSSESSIONS

Eritrea (p. 139) (85,000 square miles).
  Chief Town: Massaua.
  Products: Cattle, hides, butter.

Somali and Galla countries (p. 140) (280,000 square miles).
  Chief Station: El Adhale (Itala).
  Products: Ostrich feathers, gums.

SPANISH POSSESSIONS

Canary Islands (p. 232) (2800 square miles).
  Towns: Santa Cruz (18,000), Las Palmas.
  Products: Cochineal, oil, wheat, barley, tobacco.

Western Sahara (p. 85) (150,000 square miles).
  Station: Rio de Oro.

Fernando Po, Annobon, Corisco Bay (p. 236) (850 square miles).
  Town: (on Fernando Po) Santa Isabel (Clarence Town).
  Products: (Fernando Po) coffee, sugar, cotton, tobacco, chinchona.

BELGIAN POSSESSION

The Congo Free State (p. 220) (900,000 square miles).
  Towns and Stations: Boma, Banana, Léopoldville, Bangala, Luluaburg.
  Products: Ivory, palm oil, rubber, orchilla weed, gum copal, coffee.

THE BOER REPUBLICS

Orange Free State (p. 203) (48,000 square miles).
Towns: Bloemfontein, the capital (3400), Harrismith, Ladybrand, Kroonstad, Smithfield.
Products: Wool, hides, diamonds, ostrich feathers, coal, iron.

South African Republic (Transvaal) (p. 203) (120,000 square miles).
Towns: Pretoria, the capital (11,000), Johannesburg (50,000), Barberton, Lydenburg, Potchefstroom, Utrecht.
Products: Gold, wool, cattle, hides, ostrich feathers, coal, iron, silver.

VARIOUS

Egypt (p. 118) (? 400,000 square miles) (nominally under Turkish suzerainty, actually under British control).
Towns: Cairo (375,000), Alexandria (215,000), Damietta, Rosetta, Port Said, Suez, Tanta, Damanhur, Siut, Keneh, Assuan, Wadi Halfa, Kosseir, Suakin.
Products: Cotton, wheat, maize, rice, sugar, lentils.

The Mahdi's Dominion (p. 126) (? 600,000 square miles).
Towns: Khartum, Omdurman, Berber, El-Obeid, El-Fascher, Fashoda, Lado.
Products: Ivory, cotton, rubber, wheat, tobacco.

Abyssinia (p. 137) (150,000 square miles).
Towns: Gondar, Debra Tabor, Adua, Ankober, Antotto.
Products: Coffee, ivory, gold, hides, butter, honey, wax.

Tripolitana (p. 74) (340,000 square miles (under a Turkish Governor).
Towns: Tripoli (40,000), Benghazi, Murzuk, Ghadames.
Product: Alfa grass.

Marocco (p. 65) (150,000 square miles).
Towns: Fez (150,000), Mequinez, Marocco (50,000), Tangier, Casablanca, Mogador.
Products: Cereals, dates, hides, wool, oil, leather goods, fez caps, copper.

Wadai and Kanem (p. 98) (? 100,000 square miles).
  Chief Town: Abeshe.
  Products: Ivory, ostrich feathers.
Liberia (pp. 52, 98) (14,000 square miles).
  Towns: Monrovia (5000), Robertsport.
  Products: Coffee, palm oil and kernels, kola nuts, sugar, arrowroot, ivory.

Note.—The areas given in the above table, and in some instances the populations, are merely approximate.

INDEX

References other than Place-names are in Italics. The Numbers refer to the Pages.

ABAI, *r.*, 131
Abba Yared, Mt., 130
Abeshe, 98
Abome, 105
Abu-Simbul, 126
Abyssinia, 41, 51, 137
Abyssinian highlands, 130
Abyssinians, 21
Acacias, 17, 80, 89
Accra, 100
Adamawa, 97, 226
Addah, 100
Aden, 141
Adrar, 83
Adua, 138
Adulis, 51, 139
Afar tribe, 136
African Association, 39, 40
Agaomeder, 136
Agol, 81
Agulhas, Cape, 2
Ahaggar, 82
Ain-Sefra, 73
Air, 77, 78, 83
Akasa, 103
Akba, *r.*, 92
Akka tribe, 217
Albert Edward, Lake, 108
   Nyanza, 44, 108, 126
Alexander, Capt. J. E., 41
Alexandria, 119, 120
Alfa grass, 17, 44, 75
Alfred, Port, 198
Algeria, 61, 69, 73
Algiers, 70

Algoa Bay, 198
Alicedale, 200
Aloes, 142
Alpine flora, 132
Ambaca, 219
Ambas Bay, 226
Ambriz, 220
Amhara, 137
Andersson, C. J., 44
Angola, 219
Angoni tribe, 171
Angra, 234
   Pequena Bay, 206
Ankober, 138
Annobon, 236
Antananarivo, 230
Anti-Atlas, *m.*, 61, 62
Arabi Pasha, 54
Arabs, 29, 35, 65, 81, 117
Archil. See Orchilla
Arussi, 137
Aruwimi, *r.*, 212
Asaba, 103
Ascension, *is.*, 234
Ashanti, 52, 94, 100
Assab Bay, 57, 139
Assuan, 122
Atbara, *r.*, 109, 131
Athi, *r.*, 149
Atlas, *m.*, 5, 60
Aujila, 64, 75
Avelimmiden, 82
Awemba, 172
Axim, 100
Axum, 138

A-Zande tribe, 118, 216
Azjer tribe, 82
Azores, 233

BAB-EL-MANDEB, Straits of, 1
Baling, *r.*, 87
Bagamoyo, 161
Bagara tribe, 117
Baghirmi, 97, 98
Bagida, 106
Bahiuda steppe, 112
Bahr-el-Abiad, *r.*, 109
 el-Azrek, *r.*, 109, 131
 el-Ghazal, *r.*, 89, 109
Bahrieh Oasis, 113
Bahr Yusuf, 114
Baikie, Dr., 43
Baines, T., 44
Baker, Sir Samuel, 44
Bakhoi, *r.*, 87
Baluba tribe, 217
Bamako, 104
Bamboos, 151
Banana, 17, 151
Banana, Port, 223
Bandama, *r.*, 92
Bandiagara, 104
Bangala, 222
Bangweolo, Lake, 148, 209
Banks, Sir Joseph, 39
Bantu, 22, 153, 171, 190, 216
Banyans, 157
Baobab, 17, 170
Barbary Corsairs, 32, 69
Barberton, 205
Bardera, 137
Bariba tribe, 94
Baringo, Lake, 118
Barka plateau, 64
Barotse tribe, 58, 172
Barrage, 115
Barrakunda Falls, 87, 99
Barri, 137
Barth, Dr., 42
Bartholomew Diaz, 36
Baso, 138
Bassam, Grand, 105
Basso Narbor, Lake, 148
 Norok, Lake, 131, 147
Basutos, 191

Batanga, 226
Bathurst, 99
Batna, 70
Batua tribe, 218
Baumann, Dr. O., 48
Bechuanaland, 41, 57, 202
Bechuanas, 190, 191
Bedawin, 117
Beira, 173, 179
Beja tribes, 21, 117
Beke, Dr., 41
Benghazi, 75, 84
Benguela, 218, 220
Benin, Bight of, 90
Benti, 105
Benue, *r.*, 9, 88
Berber, 128
Berbera, 141
Berbers, 21, 65
Betsileo tribe, 229
Bihe, 208
Bilad-es-Sudan, 86
Bilma, 83
Bimbia, 226
Binger, Capt., 47
Birket-el-Kerun, 114
Bisharin tribe, 21, 117
Biskra, 70
Bismarckburg, 106
Bissagos, *is.*, 105
Bissandugu, 104
Bissao, 106
Bitter Lakes, 124
Bizerta, 74
Blanc, Cape, 62
Blanco, Cape, 85
Blantyre, 175, 179
Blidah, 70
Bloemfontein, 203
Blue River, 38, 109, 131
Bodele, 89
Boffa, 105
Boke, 105
Boma, 221
Bon, Cape, 62
Bona, 70, 72
Bonga, 139
Bongo tribe, 118
Bontuku, 95
Booruns, 137

INDEX 253

Borgu, 94
Borku, 83, 98
Bornu, 50, 97
Bosch Veld, 204
Botletli, r., 185
Böttego, Capt., 47
Bougie, 70, 72
Brava, 140
Brazza, J. de, 47
 S. de, 47
Brazzaville, 224
Browne, W. G., 39
Bruce, James, 38
Brun, André, 38
Bubi tribe, 236
Büchner, Dr., 47
Bue, 224
Bukoba, 161
Bulak, 120
Bulawayo, 179
Bulhar, 141
Burlos, Lake of, 115
Burton, Sir R. F., 44
Bushmen, 23, 30, 191

CACAO, 177, 225, 226, 236
Cacheo, r., 106
Cailland, F., 41
Caillié, R., 40
Cairo, 120
 Old, 120
Cameron, Commr. V. L., 45
Cameroons, 55, 225
 Peak, 5, 226
Canary Islands, 3, 232
Cão, Diogo, 36
Cape Coast Castle, 100
Cape Colony, 191
Capello, H. de B., 47
Cape Town, 53, 195, 196
Cape Verde Islands, 3, 234
Carthage, 74
Casablanca, 68
Casamanza, r., 91
Cassara, 93, 153
Cathcart, 200
Cecchi, Capt. A., 47
Ceratonia, 63
Cetewayo, King, 201
Ceuta, 68

Chad, Lake, 5, 8, 77, 89
Chambezi, r., 208
Chapman, J., 44
Charter, Fort, 179
Chiarini, Dr. G., 47
Chibokwe tribe, 217
Chimoio, 179
Chinde, r., 168, 173
Chobe, r., 166, 205
Chuapa, r., 211
Cidade, 236
Cirta, 70
Clapperton Capt. H., 40, 43
Clarence Peak, 236
Clores, 157
Coal, 171, 178, 195, 201
Coama, r., 168
Coanza, r., 213
Coco-nut palm, 92, 157
Coffee, 17, 98, 132, 201, 220, 223,
 225, 226, 236
Colenso, 202
Comoro Islands, 231
Compagnon, 38
Compass Berge, 183
Congo basin, 45, 168, 208
 kingdom, 55
 Railway, 222
 r., 7, 9, 39, 208
Constantine, 70, 72
Copper, 64, 72, 128, 195, 206
Copra, 157
Coriander, 63
Cotton, 17, 90, 123, 128, 177
Covilham, Pedro de, 37
Crampel, P., 47
Crocodile, r., 185
Crystal Range, 208
Cypress, 178

D'ABBADIE, Antoine and Arnauld, 41
Dahome, 94, 105
Dakhel, 113
Damanhur, 122
Damara tribes, 206
Damaraland, 41, 187, 206
Damietta, 110, 114, 122
Danákil tribes, 136
Dar-es-Salaam, 161

Darfur, 112, 126
Dar Runga, 98
Date palm, 17, 25, 81
Debra Tabor, 138
Debu, Lake, 88
Decoeur, Capt., 47
De Kaap gold-fields, 205
Delagoa Bay, 2, 169
Delcommune, A., 47
Delgado, Cape, 169
Delta of the Nile, 114, 122
Denham, Major, 40, 43
Devil's Peak, the, 196
Diamonds, 194, 199, 203
Diaz, Bartholomew, 36
Diego Suarez Bay, 230
Diele, 224
Dinka tribe, 118
Diogo Cão, 36
Draa, r., 60, 62, 66
Dragon tree, 233
Drakensberg, m., 7, 182
Dum palm, 89
Durban, 202
Durra, 25, 89, 98, 123
Duveyrier, H., 43

EASTERN HORN, 2, 129, 133
Edfu, 126
Egga, 103
Egypt, 41, 57, 115, 118
El Adhale, 141
Elaeis guiniensis, 15
El-Fascher, 128
El-Gisr, 124
El Golea, 72
El Juf, 77
El Kantara, 73
Elmina, 100. *See* San Jorge
El-Obeid, 128
Emin Pasha, 17, 127
Eratosthenes, 35
Erg or Igidi, 79
Ericaceae, 14, 187
Eritrea, 139
Erskine, St. Vincent, 44
Esueh, 122
Esparto grass, 17. *See Alfa*
Ethel, 81
Ethiopia, 129

Euphorbias, 16, 150, 233
Eyassi, Lake, 148

FALEME, r., 87
False Bay, 2
Fan tribe, 216
Farafrah, 113
Fashoda, 128
Fayum, 114, 120, 122
Fellahin, 21, 117
Fellata. See Fulbe
Felu Falls, 87
Fernando Po, 3, 236
Fez, 67
Fezzan, 74
Fianarantsoa, 230
Fig, 25
  sycamore, 116
Flax, 63
Fogo, 234
Fontesvilla, 173
Franceville, 224
François, C. von, 47
Frankincense, 136
Freetown, 99
Freretown, 159
Fulbe, Fulahs, or *Fellata*, 21, 50, 93
Funchal, 233
Fur tribe, 117
Futa Jallon, 87, 104

GABES, Gulf of, 2, 64
Gabun estuary, 53, 224
Galla country, 130, 133
Gallas, 21, 136
Galton, F., 44
Gama, Vasco da, 36, 200
Gambia, r., 38, 87, 99
Gando, 57, 95, 97
Gan Libah, m., 133
Garua, 97
Geba, r., 91, 106
Gelidi, 137
George IV. Falls, 181
Geryville, 70
Ghadames, 75
Gibraltar, Straits of, 1
Gizeh, 118, 125
Gobat, Rev. S., 41

INDEX

Gojam, 130, 131, 137
Gold, 75, 100, 139, 179, 203, 204
Gold Coast, 53, 91, 100
Goletta, 73
Gombe, 96
Gondar, 138
Gonyé Falls, 166
Good Hope, Cape of, 2, 36, 189, 191
Gordon, Gen., 54, 126, 127
Gorée, 104
Götzen, Count von, 48
Gough Island, 235
Graaf Reinet, 199
Grahamstown, 199
Grain Coast, 91
Gran Canaria, 233
Grande Rivière, 232
Grant, Col. J. A., 44
Great Aughrabies Falls, 184
Great Fish River, 185
Greeks, 35, 49
Gregory, Dr. J. W., 48
Grenfell, Rev. G., 47
Greytown, 202
Griqualand West, 193, 194
Groote Zwarte Berge, 183
Grüner, Dr., 47
Guanches, 233
Guardafui, Cape, 2, 51, 133
Guelma, 70
Guina Falls, 87
Guinea, French, 105
    Gulf of, 29, 86
    Lower, 207
    Upper, 35, 90
Gum acacias, 17, 80
    copal tree, 17, 100
Gungunyana, 171
Gurma, 97
Gurunga tribe, 94

HABESH, 136
Hadendoas, 117
Hamites, 21, 117
Hammada, 64, 80
Hanno's voyage, 31
Hanyani, r., 167
Harar, 139, 141
    m. of, 133

Harrismith, 202
Hausas, 93, 94
Hawash, r., 130, 133
Heaths, 14, 187
Hemp, 63
Henry of Portugal, Prince, 35
Herodotus, 39
Hofrat-en-Nahas, 128
Hühnel, Lieut. von, 48
Holub, Dr. E., 44
Hombori, m. of, 88
Hooge Veld, 204
Hornemann, F. C., 40
Horus, Temple of, 126
Hottentots, 23, 30, 190, 191
Humbe, 220
Humpata, 220

IBADAN, 102
Ibi, 103
Ibsambul, 126
Igharghar, Wadi, 78
Ikopa, r., 228
Ilorin, 94
Imerina, 230
Indiarubber, 17, 92, 161, 174, 215, 220, 225
Indigo, 93, 98, 132
Iron, 64, 72, 178, 195, 203
Ironwood, 188
Ismailia, 122
Itala, 141
Ivens, R., 47
Ivory, 75, 98, 139, 158, 174, 177, 223, 226
Ivory Coast, 91, 105

JALO, 64, 75
James, F. L., 47
Jamestown, 235
Jebel Aiashi, 61
    Amur, 61
    Aures, 62
    Babor, 61
    Dukhan, 113
    Elba, 113
    Gharib, 113
    Hamada, 113
    Lalla, 61
    Marrah, 112

Jebel Mokattam, 120
  Sheliya, 62
  Soturba, 113
  Um Delpha, 113
  Zebara, 113
Jenne, 104
Jibuti, 142
Jobson, 38
Johannesburg, 204
Johnston, Keith, 47
Jub, *r.*, 134
Jühlke, Dr., 160
Junker, Dr. W., 47
Jurjura, 61

KABABISH tribe, 117
Kabara, 104
Kabylia, 61
Kaffa, 135, 137
Kaffirs, 190
Kafukwe or Kafue, *r.*, 166
Kagera, *r.*, 108, 148
Kairwan, 74
Kalahari, 188
Kanem, 98
Kano, 84, 96
Kanuri tribe, 91, 97
Karagwe, 162
Karroo, 188, 195
Kasala, 128, 140
Kasongo, 222
Kassai, *r.*, 210, 222
Kassange, 218
Katanga, 222
Katsina, 96
Kauar, 83
Kayes, 104
Kebra-basa rapids, 168
Kelli, 96
Kei, *r.*, 195
Kel-Ui tribe, 82
Keneh, 122
Kenya, *m.*, 6, 141
Khalifa Abdullahi, 127
Kharghch, 113, 123
Khartum, 117, 126, 128
Khoi-Khoin, 23
Khor Baraka, 111
Kilima Njaro, *m.*, 6, 114, 162
Kilwa, 161

Kilwa Kiswani, 161
  Kivinja, 161
Kimberley, 199
Kioko tribe, 217
Kirunga, *m.*, 48, 146
Kismayu, 160
Kivu, Lake, 48, 147
Kola nut, 17, 100
Kolbe, P., 38
Konakri, 105
Konde, 162
Kong, 95, 101
  Mountains, 88
Kopjes, 203
Korata, 138
Kordofan, 127
Korosko, 122
Kosseir, 122
Kowie, *r.*, 198
Krapf, Dr., 41, 42
Kru tribe, 91
Krumir tribe, 73
Kubango, *r.*, 185
Kufra, 83, 84
Kuisip, *r.*, 185
Kuka, 97
Kumasi, 100
Kundelungu, 208
Kunene, *r.*, 213
Kwa, *r.*, 211
Kwando, *r.*, 166
Kwango, *r.*, 211
Kwanza, *r.*, 213
Kwilu, *r.*, 221

LACERDA, Dr. J. de, 40, 172
Lado, 128
Ladysmith, 202
Lagos, 53, 101
Laing, Major, 40
Lambarene, 224
Lander, 40
Landins, 171
Lange Berge, 183
Langenburg, 162
Las Palmas, 233
Lebombo range, 185
Leeward Islands, 234
Lentils, 123
Lenz, Dr. O., 43

Leopard River, 134
Leopold, King, 54
Leopold II., Lake, 211
Leopoldville, 221
Letorzek, 41
Le Vaillant, 38
Liambai, r., 165
Liba, r., 164
Liberia, 52, 98
Libreville, 224
Libyan desert, 59, 77
Licka, 139
Limpopo, r., 10, 170, 185
Lion Mountain, 196
Livingstone, David, 42, 43
Livingstone Falls, 213
    Mountains, 146
Loanda, St. Paul de, 218, 219
Loango, 224
Loangwa, r., 167
Lobelia, tree, 151
Lo Bengula, 57, 178
Lobo, Jeronymo, 37
Locust beans, 63
Logh, 137
Lokoja, 103
Lom, r., 213
Lomami, r., 211
London, East, 198
Lopez, Cape, 213
Lopez, Duarte, 36
Lopo Gonçalves, Cape, 213
Los Islands, 99
Lourenço Marques, 173, 205
Lovalé, 165
Luabo, Eastern, r., 168
Lualaba, r., 45, 209
Luapula, r., 209
Lubilash, r., 211
Lugard, Captain F. D., 47, 156
Lugh. See Logh
Lujenda, r., 149
Lukuga, r., 209
Lulongo, r., 211
Lulua, r., 211
Luluaburg, 222
Lupata Narrows, 168
Lurio, r., 169
Lydenburg, 204
Lyon, Capt. G. F., 40

MABAS, 98
Mackinnon, Sir W., 57, 155
Madagascar, 3, 36, 227
Madeira, 233
Madiville, 224
Mafeking, 179, 199
Magalies range, 204
Magdala, 138
Magdoshu, 140
Maghreb-el-Aksa, 65
Mahdi, 127
Mahé, 232
Mahebourg, 232
Mahmudiye Canal, 121
Maistre, C., 47
Maize, 25, 93, 123, 194
Majuba Hill, 202
Makololo tribe, 172
Malagarazi, r., 147
Malanje, 219
Malays, 229
Malindi, 160
Mambunda tribe, 172
Mandingos, 93
Mangbattu, 118, 216
Mangoka, r., 228
Mangroves, 90, 168
Manna, r., 99
Manyara, Lake, 148
Manyuema tribe, 217
Marakesh, 67
Mareotis or Mariut, Lake, 121
Marocco, 61, 65, 68
Masai tribe, 153
Mascarenhas or Mascarene Islands,
    227, 231
Masenya, 98
Mashonaland, 164, 178
Massari, Lieut., 47
Massaua, 57, 140
Massina, 95, 104
Matabili, 171, 178
Matabililand, 57, 164, 178
Matoppo Hills, 164
Matteucci, Dr., 47
Mauch, K., 44
Mauritius, 231
Mayotte, 231
Mehemet Ali, 41
Mejerda, r., 62, 72

S

Meknes, 67
Melakori, r., 99, 105
Melk, Wadi, 112
Memphis, 126
Menuthias, 229
Menzaleh, Lake, 115, 120
Mequinez, 67
Meroe, 126
Mfini, r., 211
Mfumbiro, m., 146
Millet, 98
Mimosas, 14
Misahöhe, 106
Mizon, Lieut., 47
Mlanji, Mt., 169
Mobangi, r., 211, 224
Moffat, Dr. R., 44, 174
Mogador, 68
Mohr tribe, 94
Mojanga, 230
Molteno, 200
Moluya, r., 62
Mombasa, 156, 158
Monbuttu, 118, 216
Monrovia, 98
Mont aux Sources, 182
Monteil, Col., 47
Mosi-oa-tunya, 166
Mossamedes, 219, 220
Mossi, 94
Mostaganem, 70, 72
Mosuril Bay, 173
Mozambique, 47, 173
    Channel, 2, 227
Msidi, 222
Mtesa, 156
Murchison Falls, 108
Muri, 96
Murzuk, 40, 43, 75
Mwanga, 156
Mwero, Lake, 58, 118
Myrrh, 136

NACHTIGAL, Dr. G., 43
Nachor, Basso, 118
Naivasha, Lake, 118, 158
Namaqualand, Little, 195
Namaquas, 41
Nanuli Hills, 169
Natal, 193, 200

Natalia, 200
Negroes, 21, 29, 86, 93
    Bantu. See Bantu
    Nilotic, 117
    Sudan, 21, 93, 111
    Swamp, 118
Nero, 35
Newcastle, 202
Ngami, Lake, 9, 42, 185
Niadi, r., 224
Niam-Niam, 118, 126
Nianam, r., 148
Nieuwveld range, 183
Niger, r., 8, 39, 57, 88, 95
Niger Coast Protectorate, 102
Niger Company, Royal, 57, 102
Nikki, 94
Nile, 8, 107
    Blue, 38, 109, 131
    Delta, 114, 122
    Somerset, 108
    White, 41, 109
Nogal, 134
Nolloth, Port, 196
Nuba tribe, 117
Nubia, 119
Nubian desert, 113
Nubians, 117
Nupe, 97
Nyangwe, 45, 209, 222
Nyasa, Lake, 44, 58, 147
Nyasaland, 57, 175
Nyika, 144
Nzoia, r., 158

OASES of Sahara, 80-83
Oasis, Great, 113
    Little, 113
Obongo tribe, 217
Ocotea, 188
Ogaden, 131
Ogowe, r., 213
Ohrwalder, Father J., 127
Oil palm, 15, 92, 215
Okavango, r., 185
Olifant, r., 185
Olive, 25
Omatako, Mt., 181
Omdurman, 128
Omo, r., 131

INDEX 259

O'Neill, H. E., 47, 173
Ongulahi, r., 228
Onitsha, 103
Ookiep, copper mines of, 195
Oran, 70, 72
Orange, r., 10, 41, 184
Orchilla lichen, 17, 161, 215
Oromo, 137
Ostrich feathers, 75, 98, 194, 203
Otyimbingue, 206
Oudney, Dr., 40
Overweg, Dr., 42
Oyo, 102

Paarl, 198, 200
Palmas, Cape, 90
Palm, coco-nut, 92, 157
   oil, 15, 92, 215
   wine, 15
Palm-oil and kernels, 100-102, 105, 106, 220, 223, 225, 226
Pandanus, 17
Pangani, 161
Panyami, r., 167
Papyrus, 116
Park, Mungo, 39
Passarge, Dr. S., 47
Payva, Affonso de, 37
Peddie, Major, 39
Pemba, is., 3, 150
Pepper Coast, 91
Peters, Dr., 160
Pfeil, Count, 160
Pharos, is., 121
Philae, 123, 126
Philippeville, 70, 72
Phoenicians, 34, 49, 65
Phoenix reclinata, 187
Pico, is., 233
Pietermaritzburg, 200
Pinto, Major Serpa, 47
Pistacia, 63
Piton des Neiges, 232
Podocarpus, 188
Pogge, Dr., 47
Ponta Delgada, 234
Popo, Great, 105
   Little, 106
Port Alfred, 198
   Elizabeth, 195, 198

Port Louis, 232
   Nolloth, 196
Porto Grande, 234
Porto Novo, 105
Port Said, 122
   Savanna, 232
   Victoria, 232
Potchefstroom, 204
Prester John, 37
Pretoria, 201
Prince's Island, 236
Ptolemy, 35, 39
Pungwe, r., 169
Pygmies, 23, 35, 217

Queenstown, 200
Quillimane, 172
Quittah, 100

Rabba, 103
Raian, the, 114
Raphia vinifera, 15
Ras Dajan, 130
Ravenala, 228
Rebmann, Dr., 42
Red Sea, 1, 12, 113, 132
Reichard, C. G., 40
Retem, 81
Réunion, is., 231, 232
Rhat, 84
Rhodesia, 177, 179
Ribeira Grande, 234
Rice, 25, 93, 123, 177, 230
Richardson, J., 42
Richmond, 202
Rio dos Camerones, 225
Rio Grande, 91, 106
Ripon Falls, 108
Ritchie, J., 40
Robecchi, Capt., 47
Rodriguez, is., 232
Rohlfs, Gerhard, 43
Romans, 35, 49
Roscher, Dr., 44
Rosebery, Fort, 177
Rosetta, 110, 114, 122
Rovuma, r., 149, 160
Rudolf, Lake, 48, 131, 143, 147
Rufiji, r., 149
Rufisque, 104

Rukwa, Lake, 146, 168
Rusizi, r., 147
Ruspoli, Prince, 47
Ruwenzori, Mt., 6, 47, 108, 146

SABAKI, r., 149, 158
Sabi, r., 170
Sahara, the, 32, 35, 42, 76
St. Denis, 232
St. Helena, 235
St. Louis, 104
St. Michael, 234
St. Paul de Loanda, 218, 219
St. Pierre, 232
St. Thomas, 236
Sakalavas, 229
Salaga, 95, 106
Saldania, A. de, 38
Saldania Bay, 38, 195
Salisbury, Fort, 179
Salt, Henry, 41
Sandwich Haven, 206
Sangha, r., 212, 224
San Jorge da Mina, 52. *See* Elmina
Sankuru, r., 211
Sannaga, r., 213
San Salvador, 218
Santa Cruz, 233
Santa Isabel, 236
Santiago, 234
Santo Antão, 234
Sanyati, r., 167
São Miguel, 234
São Vicente, 234
Saracens, 32, 49
Searcies, r., 92
Schweinfurth, Dr. G., 46, 211
Sebu, r., 62
Segu, 104
Selous, F. C., 44, 179
Semites, 21, 117, 136. *See Arabs*
Semliki, r., 108
Sena, 172
Senerio, 151
Senegal, r., 38, 58, 87, 103
Senegambia, 41
Sennar, 98
Serra do Crystal, 208, 213
Sétif, m., 61

Seychelles, 232
Sfax, 74
Shari, r., 89
Shelif, r., 70
Shilluk tribe, 118
Shiré, r., 147, 164
Shirwa, Lake, 147
Shoa, 51, 137
Shotts, 61, 62, 74
Sidi-bel-Abbes, 70
Sidra, Gulf of, 2
Sierra Leone, 59, 99
Simen, m., 130
Simon's Bay, 196
    Town, 196
Siut, 122
Siwah, 113
Slatin Pasha, 127
Slave Coast, 91
Slave trade, 26, 117, 154, 172, 175
Smith, Dr. Andrew, 41
  Dr. Donaldson, 48
Sneeuw Berge, 183
Sneezewood, 188
Sobat, r., 109
Sobso, 139
Socotrine aloes, 142
Sofala, 37, 172
Sokota, 138
Sokoto, 50, 57, 95
Sokotra, 3, 141
Somaliland, 47, 133, 141
Somalis, 21, 136
Songhay, 94
Sorghum, 89, 98. *See* Durra
Speke, Capt. J. H., 44
Sphinx, 125
Stanley, H. M., 46, 47, 127
Stanley Falls, 209, 221
    Pool, 209
Stefanie, Lake, 148
Stephanieville, 224
Stevenson Road, 175
Stinkwood, 188
Suakin, 122, 127
Sudan, 43, 50, 86
  Central, 58
  Egyptian, 126
  Nilotic, 111, 115, 126

INDEX

Suez Canal, 119, 123
  Isthmus of, 114
Sugar-cane, 123, 132, 153, 174, 177, 201, 231, 232, 236
Sus, Wadi, 62
Susa, 74
Swahili, 154
Swakop, *r.*, 185, 206
Swaziland, 204
Sycamore fig, 116
Syrtes, 2, 64
Syrtis, Little, 64

TABLE BAY, 38, 196
  Mountain, 196
Tajura Bay, 133, 142
Takazze, *r.*, 131, 137
Talha, 81
Tamarida, 142
Tamarind, 89
Tamarisk, 81
Tamatave, 230
Tamjurt, 61
Tana, *r.*, 131, 149
Tanala tribe, 229
Tanga, 161
Tanganyika, Lake, 44, 58, 147
Tangier, 68
Tanta, 122
Tea, 177, 201
Tef, 132
Teleki, Count, 48
Tel-el-Kebir, 119
Tell region, 61, 69, 72
Telremt, 61
Tenduf, 84
Teneriffe, 233
Tensift, 62
Terceira, 234
Tete, 172
Thebes, 126
Thompson, 38
Thomson, Joseph, 47
Tibesti, 77, 78, 83
Tibu tribe, 81
Tidikelt, 73
Tigré, 137, 140
Tiki-tiki tribe, 217
Timbuktu, 38, 40, 43, 103
Timsah, Lake, 124

Tioghe, *r.*, 185
Tlemcen, 70
Tobacco, 93
Togo, 106
Togoland, 106
Tongaland, 201
Toutée, Capt., 47
Traveller's tree, 228
Tripoli, 41, 42, 74, 84
Tripolitana, 64, 74
Tristan da Cunha, 235
Tsana, Lake, 131
Tsiribihina, *r.*, 228
Tsoakhaub, *r.*, 185, 206
Tuareg tribes, 21, 81
Tuat, 66, 69, 73, 84
Tuckey, Capt. J. H., 39
Tugela, *r.*, 185
Tuggurt, 72
Tuli, 179
Tumilat, Wadi, 124
Tunis, 69, 73
Turks, 50

UGANDA, 58, 154, 156
Ugogo, 150
Uitenhage, 200
Ulad-Bu-Sba, 82
  Delem, 82
  Sidi-esh-Sheikh, 82
Ulanga, *r.*, 149
Um-er-Rebia, *r.*, 62
Umvukwe range, 164
Umzila, 171
Usagara, 162
Usambara, 162

VAAL, *r.*, 184
Vangele, Capt., 47
Van Reenens Pass, 202
Veld, 188
Victoria, Ambas Bay, 226
  Falls, 166
  French Guinea, 105
  Mashonaland, 179
  Nyanza, 44, 108, 148
Villa da Praia, 234
Vine, 25, 72, 194
Vogel, Dr., 43
Volksraad, 203

Volta, r., 92
Vryburg, 199

WADAI, 98
Wadi Halfa, 122
Wagadugu, 94
Wahima, 153
Walfish Bay, 44, 185
Wambutti tribe, 217
Wanyamwezi tribe, 155
Wara, 98
Wargla, 72
Warua tribe, 217
Wassulu, 104
Watussi, 153
Watwa, 218
Waube, r., 89
Webi Shabeli, r., 134
Weenen, 202
Welle, r., 112, 211
Wheat, 72, 123, 128, 177
White River, 109
Whyda, 105
Widdringtonia Whytei, 178
Windward Islands, 234
Wine, 194
 palm, 15
Wissmann, H. von, 47
Witu, 157
Witwatersrand, 185, 204

Wolofs, 93
Wolseley, Lord, 127
Wool, 194, 203
Wosho, Mt., 130
Wuruu, 95

YAKOBA, 96
Yams, 93, 94
Yao tribe, 172
Yaunde, 226
Yellala Falls, 210
Yellow wood, 188
Yendi, 95, 106
Yola, 97, 103
Yorubas, 94, 102

ZAGAZIG, 122
Zaila, 141
Zambesi, r., 9, 163
Zande tribe, 118, 216
Zanzibar, 3, 51, 56, 150, 156
Zaria, 96
Ziziphus, 81
Zomba, 177
 Mt., 169
Zuga, r., 185
Zula, 139
Zululand, 201
Zulus, 171, 190
Zumbo, 172

THE END

www.ingramcontent.com/pod-product-compliance
Lightning Source LLC
Chambersburg PA
CBHW031943230426
43672CB00010B/2034